SURFACE TOPOLOGY, 3rd Edition

"Topology" derives from the Greek τοποσ, 'topos' (space) and 'logos' (discourse) being the study of geometrical properties and spatial relations unaffected by continuous change of shape or size of figures. The German word 'topologie' was introduced in 1847 and later anglicised to 'topology' by Professor S. Lefschetz of Princeton University, USA

"Talking of education, people have now a-days" (said he) "got a strange opinion that every thing should be taught by lectures. Now, I cannot see that lectures can do so much good as reading the books from which the lectures are taken. I know nothing that can be best taught by lectures, except where experiments are to be shewn. You may teach chymestry by lectures — You might teach making of shoes by lectures!"

James Boswell: *Life of Samuel Johnson, 1766*

About Our Authors

P.A. FIRBY

Raised in Spennymoor, County Durham, Peter Firby attended the local grammar school and then moved on to Sheffield University where he gained a first class BSc in Mathematics and then a PhD for research in general topology. In 1971 he accepted a lectureship, and eventually a senior lectureship, at Exeter University. His research publications range over general topology, lattice theory, combinatorics, graph theory, geometry, and reaction-diffusion systems. In addition, he has published a number of educational articles. In 1998 he took early retirement from Exeter University and now his time is split between teaching mathematics, his interests in arts and crafts, and the farming activities of his family.

C.F. GARDINER

Born in Offchurch, near Leamington Spa, in 1930, Cyril Gardiner was awarded scholarships, first to Leamington College, the local grammar school, and then to Birmingham University, where he studied theoretical physics under Rudolph Peierls. After graduating in 1951 he fulfilled his National Service commitment by teaching mathematics and physics in the Royal Air Force, and after leaving the RAF he became a member of a research team working on the development of guided missiles for the Royal Navy. In 1955 he returned to lecturing in pure and applied mathematics for London University degrees successively at Bournemouth College of Technology (now the University of Bournemouth) and the Northern Polytechnic (now the University of North London). During his time in London he was awarded a first class honours degree in mathematics and then carried out research in abstract algebra and number theory under the direction of Professor Albrecht Fröhlich at King's College, London. In 1961 he became a lecturer in pure mathematics at the University of Wales, Aberystwyth, and then in 1965 at Exeter University. He has published several books and articles, all on topics in pure mathematics.

Surface Topology 3rd Edition

Peter A. Firby, BSc, PhD
and
Cyril F. Gardiner, BSc, MSc, FIMA
Department of Mathematics
University of Exeter

WP
WOODHEAD
PUBLISHING

Oxford   Cambridge   Philadelphia   New Delhi

Published by Woodhead Publishing Limited,
80 High Street, Sawston, Cambridge CB22 3HJ, UK
www.woodheadpublishing.com

Woodhead Publishing, 1518 Walnut Street, Suite 1100, Philadelphia, PA 19102-3406, USA

Woodhead Publishing India Private Limited, G-2, Vardaan House, 7/28 Ansari Road, Daryaganj, New Delhi – 110002, India
www.woodheadpublishingindia.com

First published by Ellis Horwood Limited, 1991
Third edition Horwood Publishing Limited, 2001
Reprinted by Woodhead Publishing Limited, 2011

British Library Cataloguing in Publication Data
A catalogue record for this book is available from the British Library

ISBN 978-1-898563-77-8

Table of contents

Notation

D_p	dihedral group
K	Klein bottle
$K_{m,n}$	complete bipartite graph on m and n vertices
K_n	complete graph on n vertices
nM	connected sum of n copies of the surface M
$0T$	sphere
P	projective plane
S	sphere
T	torus
$\gamma(G)$	characteristic of the graph G
$\chi(M)$	Euler characteristic of the surface M
$[x]$	integer part of the real number x
$o(g)$	order of an element g in a group
$|G|$	the number of elements in the group G (the order of G)
$\pi(M)$	the fundamental group of the compact surface M
$\cong$	'is isomorphic to'

Authors' preface

It is our intention in this book to provide a simple, intuitively based, and readable introduction to geometric topology (often referred to as rubber sheet geometry) which, nevertheless, achieves significant and interesting results in both the topology of surfaces and several areas of application.

It is a proven treatment which stems from a course given over many years at Exeter University to mathematicians who do not intend to specialize in algebraic topology. Nevertheless, the book should also be useful in *motivating* intending specialists in algebraic topology. Moreover, the approach is such that education students, sixth-formers, and all those keen amateurs who are avid readers of Martin Gardner's books on recreational mathematics should find it accessible.

In the usual introductory textbooks on geometric or algebraic topology the figures to be studied are built from certain finite 'bits' of space called simplexes. This is the so-called combinatorial approach. In this book, we shall follow a related approach which is adequate for dimensions ≤3. However, in the standard formal development, certain sophisticated techniques involving group theory and homological algebra are usually introduced. We shall avoid this by developing a significant theory by elementary means, in the hope that the non-specialist will thereby gain more easily an understanding and an enjoyment of this beautiful branch of contemporary mathematics.

Nevertheless in Chapter 10 we do introduce the fundamental group. This leads in a natural way to important branches of mathematics such as knot theory and combinatorial group theory. For those interested in this line of development, our book serves as a very gentle introduction to such works as *A First Course in Algebraic Topology* by Kosniowski (CUP 1980), *Classical Topology and Combinatorial Group Theory* by Stillwell (Springer 1980), and *Introduction to Knot Theory* by Crowell & Fox (Springer 1963).

For the student wishing to specialize in the other main line of thought in algebraic topology, namely homology theory, we recommend *A Combinatorial Introduction to Topology* by Henle (Freeman 1979), and *Graphs, Surfaces and Homology* by Giblin (Chapman & Hall 1977) as suitable texts for the next stage in their development.

One of the features of this subject is the large number of important topics to which it relates. Thus the text also provides a gentle introduction to graph theory, group theory, vector field theory, hyperbolic and Euclidean plane geometry, and plane tessellations. For those readers wishing to extend their knowledge of these topics, in addition to the references given in the text, we also recommend the books listed at the end of the bibliography.

Acknowledgements
Over the years many Exeter University students have taken the topology course on which this book is based, and all have therefore played a part in its final form. However, we would particularly like to acknowledge the efforts of Claire Rowland who, in a project under the supervision of Peter Firby, worked on the topics covered in Chapter 9 and contributed several interesting ideas of her own.

Preface to this 3rd edition

The wide interest attracted by our book has encouraged us to bring out a third edition. Because of the main changes of education in the United Kingdom, there is a growing demand in secondary and higher education for the type of mathematics course presented here, which opens up the area to a wider audience including intending specialists in general or algebraic topology.

With two new chapters on surfaces with boundaries and graphs and groups, the book now provides an updated, straightforward treatment of surface topology in this area which is particularly important for its richness of applications and variety of interactions with other branches of mathematics, e.g. surface topology, graph theory, group theory, vector field theory, plane nucledian and non-nucledian geometry, and knot theory. Each topic is treated from its beginnings with sufficient theory developed by elementary means, providing understanding and enjoyment.

We particularly ackowledge the helpful remarks of Professors Stewart, Griffiths and Grünbaum.

Exeter, March, 2001

1

Intuitive ideas

1.1 INTRODUCTION

Topology is an abstraction of certain geometrical ideas such as 'continuity' and 'closeness'. In fact the word 'topology' is derived from the Greek $\tau o \pi o \sigma$, a place, and $\lambda o \gamma o \sigma$, a discourse. It was introduced in 1847 by Listing, a student of Gauss, in the title of the first book devoted to the subject. Another name used in the early days was '*analysis situs*', analysis of position.

A popular term used to convey the more intuitive aspects of the subject is 'rubber-sheet' geometry. Think of those properties of figures drawn on a sheet of rubber which are not altered when the sheet is distorted. For example, intersecting curves still intersect after distortion; a circle remains a closed curve, with an inside and an outside, and so on.

Two essentially distinct developments of topology must be distinguished. Point-set topology, or general topology, is a general abstract theory of continuous functions defined on very general sets. It was influenced by the general theory of sets developed by Cantor around 1880, but received its main impetus from the theory of metric spaces (that is, abstract spaces with a 'distance' defined between points) introduced by Frechet in 1906, and from the publication of the book *Grundzüge der Mengenlehre* by Hausdorff in 1912. In the latter, Hausdorff extended the concepts of limit and continuity from sets of real numbers to abstract sets by means of the idea of the 'neighbourhood' of a point: this latter in some sense capturing the intuitive notion of 'nearness' to a point.

Parallel to this line of development, and in fact predating it by more than a decade, was Poincaré's introduction during the years 1895–1901 of the systematic study of algebraic topology, or *analysis situs*, as he called it. This was motivated by certain geometric problems about paths and surfaces in Euclidean space. The basic method of algebraic topology is to associate a group, or a sequence of groups, with the topological space in the hope that the geometrical and topological properties of the space will be reflected in the structure and properties of the associated group or groups.

Like Poincaré's, our motivation in this book is the solution of certain problems in the topology of surfaces. However, the normal approach to surface topology builds on a basic foundation of general topology. Unfortunately this involves much hard work and not a little tedium before reaching the kind of results that interest us. We avoid it here.

In this chapter we shall gather together in an intuitive form the main ideas we shall require in our investigation.

In the later chapters we shall develop definitions and techniques that will enable us to prove some of the important theorems and to discover, quickly and easily, many of the interesting properties of surfaces.

1.2 PRELIMINARY SKIRMISH

From the time we were born we have explored the world around us by means of our senses. In particular we all have an intuitive understanding of what a surface is. Unfortunately, when we come to ask precise questions about surfaces, our intuitive understanding is often too vague to give convincing answers.

Our first task, therefore, is to seek a precise definition of surface that can be used in logical argument.

Let us picture the surface of a rubber ball. Take any point on the surface and consider a small disc on the surface with this point at the centre. Now imagine that we cut this disc, of negligible thickness, out of the ball. With a little distortion, which is allowed in our 'rubber-sheet' geometry, we can turn it into a flat disc,, that is, a 2-dimensional disc lying in a plane.

We describe this situation by saying that the immediate neighbourhood of each point on the surface of the ball is 'flat'.

Notice that this latter use of the word 'flat' is *not* the ordinary usage of the word. Thus in topological language **flat** means 'can be *made* flat in the ordinary sense'.

Each point on the surface of the ball has this property of being flat in the topological sense.

A set of points which behaves in this way we call a **locally flat** set of points. Thus in particular, the surface of the ball is a locally flat set of points.

Let us state this more formally as:

DEFINITION 1.2.1   A set of points in space is **locally flat** if, over each point p in the set, we can place a small plane disc (which we call a **test disc**) so that each point of the test disc lies above a point in the set. Moreover, each point in the set in the immediate neighbourhood of p lies beneath a point in the test disc.

Fig. 1.2.1 may help to picture the situation.

We assert that the property of being locally flat is possessed by those objects we intuitively regard as surfaces.

To test our definition let us now consider the two sets of Fig. 1.2.2.

The first example consists of all the points inside a plane disc D. This set is locally flat according to our definition. Test discs placed at typical points a, b, and c, satisfy the conditions of Definition 1.2.1.

However, the set of points represented by the curve AB is not locally flat. The neighbourhoods of a typical point p in this set are 1-dimensional. Thus any disc lying

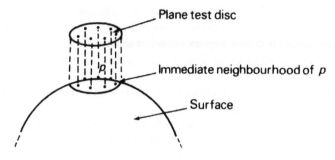

Plane test disc

Immediate neighbourhood of p

Surface

Fig. 1.2.1.

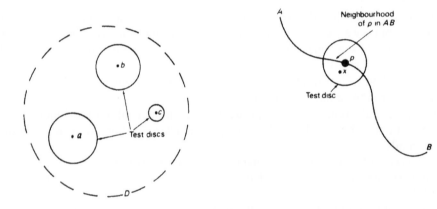

Fig. 1.2.2.

above (covering) p will contain points like x which fail to lie above (cover) a point in the set.

So far we have been thinking of our locally flat sets as sitting in the 3-dimensional space in which we live. However, the sets of Fig. 1.2.2 could equally well be considered to be part of the 2-dimensional space of the plane sheet of paper on which they are drawn.

If we imagine ourselves to be 2-dimensional people crawling about in this plane with our test discs, we see that our definition leads to exactly the same conclusions as before about the nature of the two sets in Fig. 1.2.2.

Similarly, nothing changes if we consider ourselves to be examining the same sets in 4-dimensional space, or even higher dimensions.

This leads us to the conclusion that *local flatness is a property of the sets themselves* and is independent of the space in which they are situated.

A model, or 'picture', of a set in a space of sufficient dimension to make the picture' seem natural (i.e. the set does not appear to be unnaturally restricted within its surrounding space) is called a **space model** of the set.

For example, the pictures of Fig. 1.2.2 are space models in 2 (or higher)-dimensional space.

What we see of the surface of a telephone dial is a space model of a locally flat set in 3 (or higher)-dimensional space.

The space models of many of the surfaces which concern us sit naturally in 3 (or higher)-dimensional space. But soon we shall meet surfaces that do *not* sit naturally in 3-dimensional space. They appear to be unduly restricted and cry out to be represented in a higher-dimensional space.

To conclude this section, we remark that the technical term usually used for a 'locally flat set' is '**two-dimensional manifold**'.

1.3 MODELS

To the senses of touch and sight, the surface of our rubber ball appears smooth and locally flat, but if we were to examine the surface through a microscope of sufficient power, we would see jagged regions of mountains and craters difficult to reconcile with the image formed by our senses.

It is this latter *ideal* model in our mind, arising from our unaided senses, that corresponds to the actual space model of our theory.

Later we shall create some real, usually paper, models of our ideal models. We then pretend that these real models actually are exact realizations of the ideal models in our minds.

Let us consider an important convention that we must use.

First we ask: *Is the surface of a rectangular sheet of paper a 2-dimensional manifold?*

Taking note of the 'thickness' of the paper and regarding it as a solid object in 3 dimensions, the answer is *yes*.

In fact the surface of any solid object is a 2-dimensional manifold.

Fig. 1.3.1 shows an enlarged portion of a part of the surface of the paper, giving the neighbourhoods of typical points *a*, *b*, and *c* on the surface.

Above these neighbourhoods we can imagine test discs placed to satisfy the conditions of Definition 1.2.1. One such test disc is shown.

The ideal model we carry in our minds of the surface of this sheet of paper is similar to our ideal model of the surface of a rectangular block of wood.

We now pose our second question.

Is a plane rectangle (i.e. no 'thickness') a 2-dimensional manifold?

This time the answer is *no*.

Let us look at Fig. 1.3.2. *Note* that the plane rectangle is viewed from *both* sides.

If we place a test disc over the point *p* on the boundary of the rectangle, there will always be a point like *x* in the test disc, which fails to cover a point in the rectangle.

If we take a rectangular sheet of paper as a *real* model of an *ideal* plane rectangle, we must adjust our way of seeing the paper so that it suggests to us the ideal model of Fig. 1.3.2 and not that of Fig. 1.3.1.

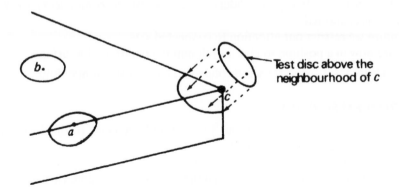

Fig. 1.3.1.

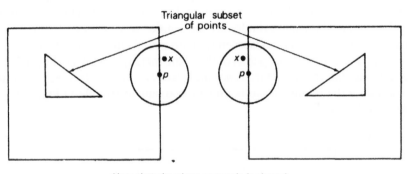

Note that the plane rectangle is viewed
from *both* sides

Fig. 1.3.2.

To do this we pretend that the paper has *no* thickness and has just *one* 'side'. We picture lines or discs drawn on the paper to be part of the paper. Thus the set of points forming a triangle in Fig. 1.3.2 is considered to be a subset of the set of points forming the rectangle and to be visible when we view the rectangle from *either* side. Hence the two views shown in Fig. 1.3.2.

1.4 CONNECTED SETS

As we have discovered, the surface of a rubber ball is a 2-dimensional manifold. But so also is the set of points forming the surface of two or more rubber balls. However, there is nothing we can learn about the surface of two or more rubber balls which we cannot deduce by putting together the information gained by examining the surface of each ball separately.

DEFINITION 1.4.1   If a space model of a set has just *one component part*, we say that the set is **connected**.

In future we restrict our attention to connected sets.

We are now in a position to give a preliminary definition of a surface.

DEFINITION 1.4.2   A **surface** is a connected 2-dimensional manifold.

1.5 PROBLEM SURFACES

Unfortunately some surfaces turn out to be trouble-makers. In this introductory book it is advantageous to omit them.

Let us consider how difficulties may arise.

Sets which are too big, in the sense that they cannot be contained in a finite space, are obvious trouble-makers.

In such a set we can travel along a path for ever, always covering 'new ground' while remaining in the set. We shall confine ourselves to the kind of set whose space model may be placed in a *finite* 'box' in the appropriate space.

DEFINITION 1.5.1   A set with this latter property is said to be **bounded**.

The surface of any solid object in everyday life is a bounded set in 3-dimensional space. The *infinite* plane and cylinder are **unbounded** sets.

However, even bounded surfaces may cause trouble in a similar way.

Consider the set of interior points of the plane rectangle shown in Fig. 1.5.1. The dashes indicate that the boundary is *not* included. Notice that this set is a bounded surface because the points on the boundary that caused trouble in Fig. 1.3.2 have now been excluded.

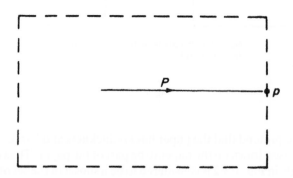

Fig. 1.5.1.

We can travel along the path *P* getting nearer and nearer to the point *p* (on the boundary) without ever being able to reach *p*, because *p*, being a boundary point, is *not* in our set.

If we enlarge our set to include *p* and the rest of the boundary of the rectangle, we avoid the difficulty of paths that go on for ever, but at the expense of returning to the object of Fig. 1.3.2, which we found is *not* a 2-dimensional manifold.

Our trouble stems from the fact that the plane rectangle has a 1-dimensional boundary.

Thus we exclude from our considerations all sets whose space models have a 1-dimensional boundary, whether included in the set or not.

In Chapter 3 we shall obtain a more precise definition of the kind of surface in which we are interested. For the moment, let us call it a **compact** surface.

It is comforting to note that we have *not* excluded any of the surfaces of the solid objects met in everyday life. All the surfaces excluded from consideration belong only to our imagination, e.g. infinite planes and cylinders, and the interiors of plane rectangles.

Nevertheless, as we shall see later, some of the surfaces of the kind we consider exist only in our minds and cannot be realized as the surfaces of real objects in 3-dimensional space.

1.6 HOMEOMORPHIC SURFACES

We have decided that the objects for study are to be bounded surfaces without 1-dimensional boundaries, i.e. compact surfaces. But in what properties of these surfaces are we interested, and how do we intend to study them?

As outlined in the introduction, we are concerned with topological properties, i.e. those properties unaltered by certain distortions of the surface.

This suggests the first method of approaching a topologist's viewpoint.

We examine surfaces as if we are handling flexible space models in 3-dimensional, or, if necessary, higher-dimensional space. We consider two surfaces to be the same and say they are **homeomorphic** if one of the space models can be *continuously* distorted to look like the other.

By continuous distortion we mean bending, stretching, and squashing without tearing or 'gluing' points together.

According to these criteria, we see that the surfaces of the balls, the cigar, and the spectacles of Fig. 1.6.1 are homeomorphic.

Similarly the surfaces of the torus and the teacup illustrated in Fig. 1.6.2 are homeomorphic.

In Fig. 1.6.3 we show a few intermediate stages in the continuous process of distortion that takes the torus into the teacup.

By examining the surfaces of the torus and the knot shown in Fig. 1.6.2 in 4-dimensional space, we find that they are homeomorphic. The process of distortion is illustrated in Fig. 1.6.4.

If we examine space models of one of the balls in Fig. 1.6.1 and and the torus in Fig. 1.6.2, we feel that we could not remove the hole in the torus in order to obtain the ball without 'gluing' together certain points.

Hence we feel that the torus and the ball are *not* homeomorphic.

This becomes more apparent when we use the second method of approach to the topologist's viewpoint. It is based on the concept of 'local flatness' discussed earlier.

We consider the surface itself to be the only object that exists for us, and imagine ourselves to be myopic 2-dimensional people living in a 2-dimensional (locally flat)

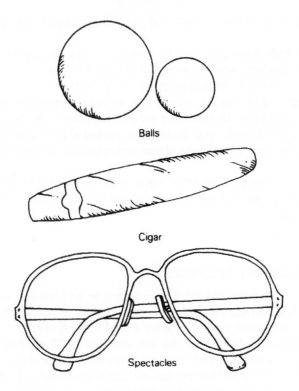

Balls

Cigar

Spectacles

Fig. 1.6.1.

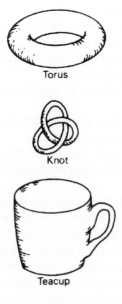

Torus

Knot

Teacup

Fig. 1.6.2.

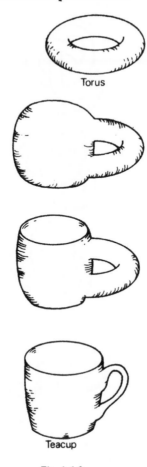

Torus

Teacup

Fig. 1.6.3.

world which is the surface being studied. We move about within this surface observing and deducing.

As 2-dimensional topologists, we draw lines on the space models of the surfaces and make some simple deductions. These convince us that the ball (or sphere) and the torus are not homeomorphic.

We illustrate the process in Fig. 1.6.5.

If we draw loop 1 on the torus, we can still move to any point on the surface without crossing the line. On the other hand, wherever we draw a loop on the sphere, we always cut ourselves off from a part of the surface.

Thus the sphere and torus do not appear to behave in the same way to us as 2-dimensional topologists. We conclude that they are *not* homeomorphic.

In this book we study surfaces from the topological point of view, so we shall not distinguish between homeomorphic surfaces. As far as we are concerned, they are all the same.

Thus when we study the topological properties of a sphere, we are at the same time studying the topological properties of the surface of a pencil, the surface of a

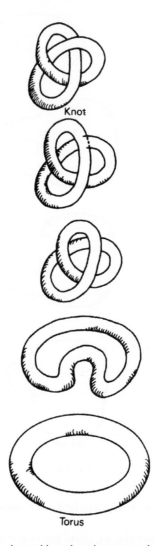

Fig. 1.6.4 — Distortions taking a knot into a torus in 4 dimensions.

pair of spectacles, the surface of the earth, and the surface of most trees when the root system is included.

In studying a ten-holed torus we are at the same time studying the surface of a telephone dial, and the surface of many examples of modern sculpture.

1.7   SOME BASIC SURFACES

There are four basic surfaces, so called because they play a fundamental role in the theory which follows. Naturally each of the basic surfaces is a compact surface.

The first is the sphere denoted by S and the second is the torus denoted by T.

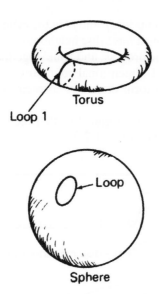

Fig. 1.6.5.

To obtain the third basic surface, let us consider what happens when we cut a torus as shown in Fig. 1.7.1(a) and (b).

We end up with the cylinder shown in Fig. 1.7.1(c).

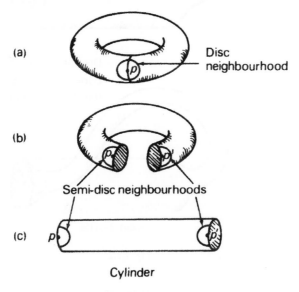

Fig. 1.7.1.

As we know from our earlier discussion of the plane rectangle in section 1.3, the existence of points p and p' on its 1-dimensional boundary prevent this set from being a 2-dimensional manifold.

The neighbourhoods of points like p and p' in the cylinder are *semi*-discs whereas a surface, being locally flat, must have disc neighbourhoods at *every* point.

Of course, if we reverse the process, and 'glue' together each corresponding pair of points like p and p' on the 1-dimensional boundaries, then we 'glue' together the semi-disc neighbourhoods to form discs and so recover the torus with which we began.

However, there is *another way* to 'glue' together these boundaries so as to obtain a surface. The process is shown in Fig. 1.7.2.

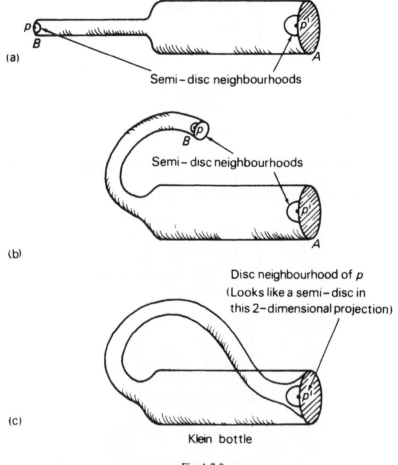

Fig. 1.7.2.

This time we bring the two 1-dimensional boundaries A and B together so that the boundary B (say) approaches and joins the boundary A from the left as illustrated in Fig. 1.7.2. Then just before making the join, the semi-disc neighbourhood of each point p is twisted before being 'glued' to the semi-disc neighbourhood of the corresponding point p', so that the required disc neighbourhoods are formed.

The resulting surface is called a **Klein bottle** and is denoted by K. This is the third of our basic surfaces.

As we can see from Fig. 1.7.2 and the difficulty of visualizing the above construction, the Klein bottle does *not* sit naturally in 2- or 3-dimensional space, where it appears to cut into itself.

To obtain space models of K, we must go into 4- or higher-dimensional space where there is more 'room' for the twisting and turning necessary in their construction.

Fig. 1.7.2(c) is a **projection** of such a 4-dimensional space model into 2-dimensional space.

The remaining basic surface we shall meet for the first time in Chapter 2.

1.8 ORIENTABILITY

As 2-dimensional topologists moving within the surfaces of the space models of Fig. 1.6.2, we cannot tell the difference between the knot and the torus. In fact, as we discovered in section 1.6, by placing the knot in 4-dimensional rather than 3-dimensional space, it becomes '*un-knotted*'.

Thus the question arises whether, as topologists, we can actually tell the difference between a torus and a Klein bottle. Perhaps by placing the Klein bottle in a space of sufficiently high dimension, it, too, might become a torus.

As fairly intelligent 2-dimensional topologists, we can mark arrows to indicate direction along the lines we draw on the space models of our surfaces.

In this way we can distinguish between clockwise and anti-clockwise. This turns out to be sufficient to distinguish between the torus T and the Klein bottle K, as can be seen from Fig. 1.8.1.

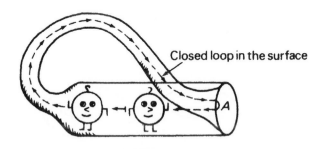

Closed loop in the surface

Fig. 1.8.1.

Imagine the little man moving within the surface along the curve shown in the direction of the arrows, involving a 'flip over' *in our diagram* as he moves through the point A. He returns to his starting point as the *mirror image* of himself.

This does *not* happen with the torus.

DEFINITION 1.8.1 If, by following some closed loop on a surface, we can change clockwise rotations $\circlearrowright$ into anti-clockwise rotations $\circlearrowleft$, we say that the surface is **non-orientable**. Otherwise we say that the surface is **orientable**.

Fig. 1.8.1 encourages us to believe that to produce a space model of a *non-*orientable surface needs a twist which cannot be handled in 3-dimensional space.

In fact this view is correct.

Thus the surface of every solid object we see about us is a bounded orientable surface with no 1-dimensional boundaries, i.e. an orientable compact surface.

1.9   THE CONNECTED SUM CONSTRUCTION

There is a simple way in which we can construct new surfaces from other surfaces.

Let p_1 be a point on the surface S_1 and let p_2 be a point on the surface S_2.

Since S_1 is a surface, it is locally flat at p_1. Thus there exists a small disc neighbourhood D_1 of p_1 which lies in the surface S_1. Similarly there exists a disc neighbourhood D_2 of p_2 in the surface S_2.

If we cut out the interiors of the discs, the sets that are left fail to be surfaces, because they have 1-dimensional boundaries; namely the perimeters of the discs.

However, by 'gluing' these boundaries together, point by point, the 1-dimensional boundaries are removed and a new surface is formed.

The surface constructed in this way from the original surfaces S_1 and S_2 is called the **connected sum** of S_1 and S_2 and is denoted by $S_1 \# S_2$.

Fig. 1.9.1 illustrates this process and shows that the connected sum of two tori is a 2-holed torus denoted by $2T$.

In like fashion the connected sum of $2T$ and T is a 3-holed torus $3T$.

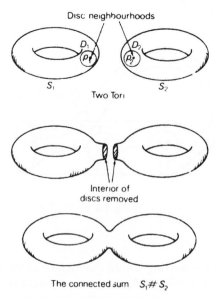

Disc neighbourhoods

D_1 D_2

p_1 p_2

S_1 S_2

Two Tori

Interior of
discs removed

The connected sum $S_1 \# S_2$

Fig. 1.9.1.

In general, by continuing in this way, we can construct an n-holed torus nT for each positive integer n.

We have the following:

THEOREM 1.9.1   An orientable compact surface is homeomorphic to either a sphere S or nT for some positive integer n.

Thus the orientable compact surfaces are precisely the surfaces of the solid objects we see about us in 3-dimensional space.

1.10 SUMMARY

We have given a preliminary definition of compact surfaces which, although not yet suitable for use in proving theorems, nevertheless helps to sharpen our intuitive ideas.

We have discussed the kind of properties of surfaces that interest us as topologists. In particular two surfaces are to be considered the same if they are homeomorphic. Here again our definition is given in largely intuitive terms.

Some basic surfaces have been listed, and we have considered the notion of orientability. This latter appears to distinguish between surfaces whose space models sit naturally in 3-dimensional space ('real' surfaces) and surfaces whose space models require 4- or higher-dimensional space.

Finally we have described the connected sum construction by means of which we can build every 'real' surface from two of the basic surfaces, namely the sphere and the torus, using these repeatedly as required.

1.11 EXERCISES

(1) Sketch the following sets.

   Which of these sets are   (i) 2-dimensional manifolds

                             (ii) connected

                             (iii) bounded

                             (v) compact surfaces.

   (a) $\{(x,y,z):x^2 + y^2 + z^2 = 1\}$

   (b) $\{(x,y,z):x^2 + y^2 + z^2 = 1 \text{ and } z \leqslant 1/2\}$

   (c) $\{(x,y,z):x^2 + y^2 + z^2 = 1 \text{ and } z < 1/2\}$

   (d) $\{(x,y):xy > 0\}$

   (e) $\{(x,y,z):x^2 + y^2 = (2 \pm \sqrt{1 - z^2})^2\}$

   (f) $\{(x,y,z):|x| + |y| + |z| \leqslant 1\}$

(2) Is the set illustrated in Fig. 1.11.1 a compact surface if

   (a) the figure represents a portion of two intersecting planes

   (b) the figure represents the 'surface' of a four-page booklet?

(3) Sketch a space model of a torus and on this describe two loop cuts which in 3-dimensional space would produce two interlocked bands.

   Would the bands be interlocked in 4-dimensional space?

(4) A box in 2-dimensional space is a square, and a box in 3-dimensional space is a cube. Sketch a Klein bottle in a box in 4-dimensional space, thus demonstrating the fact that K is bounded.

(5) Imagine the letters of the alphabet to be carved, as capitals, in wood. List the letters with surface homeomorphic to
    (a) a sphere
    (b) a torus
    (c) a 2-holded torus.

(6) By means of a sequence of sketches, as in Fig. 1.9.1, illustrate
    (a) $T\#S$ is homeomorphic to T
    (b) $S\#M$ is homeomorphic to M, for *any* surface M.

(7) Sketch the compact surface $T\#K$.

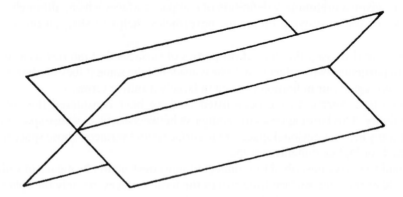

Fig. 1.11.1.

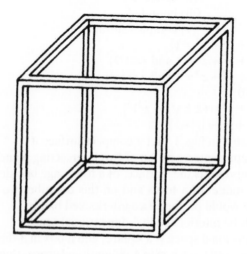

Fig. 1.11.2.

(8) (a) Construct paper models for T and K using the directions in Fig. 2.2.1. Before taping the models mark on your paper the points p, p' of Figs 1.7.1 and 1.7.2, and the semicircles representing their semi-disc neighbourhoods. Examine the finished models to see how these semi-discs fit together to form a disc neighbourhood of the identified point p.

(b) Sketch a sequence of clockwise arrows on your model for K, using Fig. 1.8.1 as a guide, to show that K is *non*-orientable.

(c) What do you get when you cut your paper model of K in half along the line l shown in Fig. 2.2.1?

(d) How could you cut a paper model of K with a *single* loop cut to produce a *single* Möbius band?

(9) Instead of taking a disc as a basic 2-dimensional neighbourhood of a point, take a line segment as a basic 1-dimensional neighbourhood of a point and mimic the 2-dimensional definitions to describe a '1-dimensional manifold', a '1-dimensional surface', and a 'bounded 1-dimensional surface with *no* zero-dimensional boundaries', sketching examples to illustrate the differences.

(10) Theorem 1.9.1 tells us that the surface of the wooden framework for a box shown in Fig. 1.11.2 can be expressed as a connected sum of a certain number of tori. How many tori?

2

Plane models of surfaces

2.1 THE BASIC PLANE MODELS

While space models do actually represent what we 'see' of surfaces, they can be difficult to sketch and are certainly cumbersome when used to help solve problems. However, by 'cutting up' space models, we can produce simple polygonal representations of surfaces which do not suffer from these drawbacks. In fact, such polygons turn out to be so useful in practice that in the next chapter we base our working definition of a surface on them.

DEFINITION 2.1.1   A polygonal representation of a surface is called a **plane model** of the surface.

Fig 2.1.1 shows how we can derive the plane model of a torus.

We make two loop cuts l_1 and l_2 as shown in the diagram. These allow us to open out the torus to form a plane rectangle.

By section 1.3, we know that the plane rectangle with its 1-dimensional boundary is not a surface, but, by reversing the cutting process just described and 'gluing' together (identifying) opposite edges of the rectangle, we can eliminate this boundary and reconstruct the torus.

By *labelling* the edges it is possible to include the rectangle and the construction in a *single* diagram, as shown in Fig. 2.1.2.

We interpret this diagram to mean that, when constructing the space model of the surface, first we identify the edges a, ensuring that the arrows point in the same direction, thus obtaining a cylinder, then we identify the edges b, which now form the circular boundaries of the cylinder, respecting the directions of the arrows as before.

Thus the points x in Fig. 2.1.2 represent a *single point*, and the oblique lines represent a *single loop* on the torus T.

It will prove to be convenient to add a further refinement to our models by making the labels attached to the edges carry the information given by the arrows.

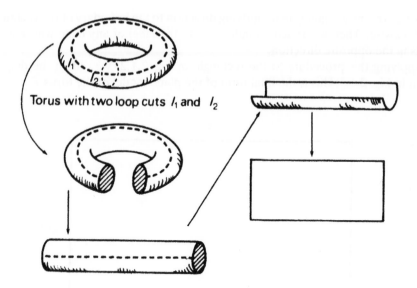

Torus with two loop cuts l_1 and l_2

Fig. 2.1.1.

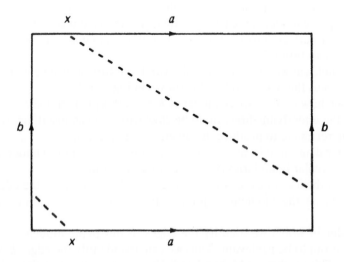

Fig. 2.1.2.

We do this by assigning an underlying direction to the boundary of the rectangle; say clockwise. Then we attach an index -1 to the label of any edge whose arrow points in the opposite direction.

Applying this procedure to the rectangle of Fig. 2.1.2, we obtain the diagram shown in Fig. 2.1.3 This is the final form of the plane model of the torus T.

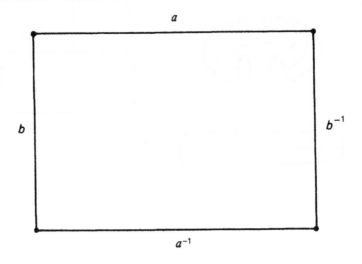

Fig. 2.1.3.

Notice that the vertices of this rectangle represent a *single* point on the torus.

To find a plane model of the Klein bottle K, recall that our construction of a space model of this surface involved identifying the 1-dimensional boundaries of a cylinder, as in section 1.7.

With a single cut we can open out this cylinder to form a rectangle. Then suitable labelling gives us the plane model of K shown in Fig. 2.1.4.

Let us see how to get a space model of K from this plane model.

Taking the underlying direction to be clockwise, we can insert arrows along the edges of the rectangle to produce the diagram in Fig. 2.1.5.

This latter diagram tells us to identify the edges a to form a cylinder and then to identify the circular boundaries b of the cylinder to form K.

In each of the identifications we must respect the arrows, i.e. the edges must be joined so that at the moment of joining the arrows involved point in the same direction.

The order in which we identify the edges of our plane models to form space models turns out to be irrelevant. Thus we can first identify the edges b of the plane model of K. This produces a **Möbius band**. Then we can eliminate the 1-dimensional edge of this Möbius band by identifying the edges labelled a in the original rectangle. Again a Klein bottle is formed. Try to sketch this procedure!

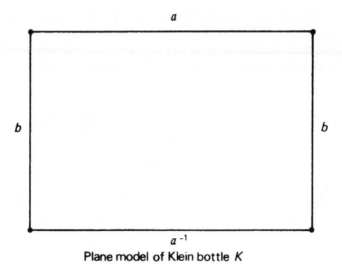

Plane model of Klein bottle *K*

Fig. 2.1.4.

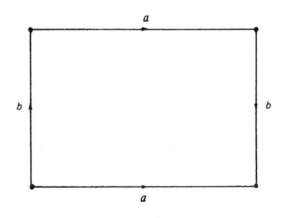

Fig. 2.1.5.

Notice that the vertices of the plane model of *K* represent a single point on the space model.

Let us apply this technique to the third of our basic surfaces, the sphere.

As we see in Fig. 2.1.6, a single cut in a sphere is sufficient to form a plane model. It is a 2-gon with suitably labelled edges.

Notice that the vertices of this plane model for *S* represent *two* points on the space model.

We are now ready to construct the fourth and final basic surface referred to in section 1.7 but not constructed there.

In 3-dimensional space the natural way to remove the 1-dimensional boundaries of a 2-gon by identification is given in Fig. 2.1.6. This gives a sphere.

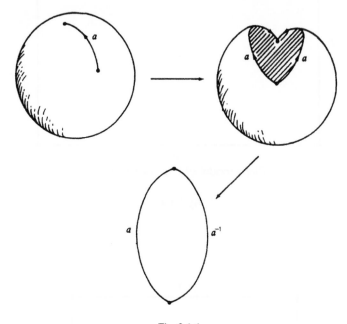

Fig. 2.1.6.

However there is another way.

We twist the figure and 'glue' one edge to the other edge 'backwards'. This is not easy to imagine in 3-dimensional space. The procedure is represented by the plane model of Fig. 2.1.7(a).

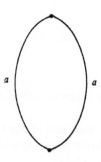

Fig. 2.1.7(a).

As we can see from Fig. 2.1.7(b), it amounts to identifying diametrically opposite points of a disc.

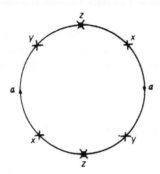

Fig. 2.1.7(b).

We call our fourth basic surface the **projective plane** and denote it by P.

Fig. 2.1.8 illustrates the construction of a space model of P from its plane model.

Notice that the vertices of the plane 2-gon model of P represent a *single* point on the space model.

Since the construction of a space model from a plane model involves identifying edges *in pairs* to remove all 1-dimensional boundaries, only polygons with an *even* number of edges can occur as plane models of surfaces.

As we see in Fig. 2.1.9, the plane model of a surface is *not* unique. In fact, we can equally well represent a sphere by *any* 2n-gon with suitably labelled edges, where n is a positive integer.

Naturally we normally use polygons with the *least* number of edges sufficient for the purpose.

To a topologist, 2n-gons and discs are essentially the same, i.e. homeomorphic.

In future we shall take the plane models to be discs with vertices and labelled arcs on the boundaries.

This will make the task of drawing the plane models much easier.

Our final list of basic surfaces, with the plane models drawn as discs, is shown in Fig. 2.1.10.

2.2 PAPER MODELS OF THE BASIC SURFACES

It is simple to describe a method of obtaining paper models of the basic surfaces from the corresponding plane models.

First take a sheet of paper to represent the basic polygon of the plane model. Then tape together the edges that have to be identified, paying due regard to the directions of the edges.

However, since we are doing this in 3-dimensional space, it is necessary to cut the paper to allow part of a *non*-orientable surface to pass through another part of the surface. In the final model we pretend such cuts are not there. In fact we have to imagine that the paper model is in 4 or higher dimensions, where the space model that it represents actually exists. The paper model is a realization of the projection of the space model into 3 dimensions.

We must remember the discussion in Chapter 1 when working with these models. In particular we must pretend that the paper is not opaque and has no thickness.

Instructions for the construction of paper models of the basic surfaces from paper polygons are given in Figs 2.2.1 and 2.2.2.

2.3   PLANE MODELS AND ORIENTABILITY

The path taken by the 2-dimensional topologists in Fig. 1.8.1 when demonstrating the *non*-orientability of K may be described on the plane model of K as illustrated in Fig. 2.3.1.

This demonstration shows that K is *non*-orientable essentially because its surface contains a Möbius band along which the path is taken: see Fig. 2.3.2.

In the light of this insight, let us consider the orientability of our fourth basic surface, the projective plane.

As shown in Fig. 2.3.3, it is simple to trace out a Möbius band on a plane model for P. It follows that the projective plane is also a *non*-orientable compact surface.

Of course we suspected this from the difficulty of imagining the projective plane in 3-dimensional space when it was constructed in section 2.1.

The space model of P exists naturally in 4-dimensional space, but not in space of lower dimension.

2.4   CONNECTED SUMS OF THE BASIC SURFACES

We have already described in Theorem 1.9.1 every compact surface that we can picture in 3-dimensional space, i.e. whose space model exists in 3-dimensional space.

Now that we have at out disposal the two basic *non*-orientable compact surfaces, we can extend this theorem to describe every imaginable compact surface.

For convenience, from now on let us adopt the convention that $0T$ (a torus with no holes!) is homeomorphic to S.

Formally we have:

THEOREM 2.4.1   An orientable compact surface is homeomorphic to nT, for some $n \geq 0$. A *non*-orientable compact surface is homeomorphic to

$$
\begin{array}{ll}
\text{either} & nT\#K \\
\text{or} & nT\#P, \\
\text{for some} & n \geq 0.
\end{array}
$$

2.5   SUMMARY

We have met the fourth basic surface, the projective plane, and we have seen that each basic surface can be represented by a 2n-gon with suitably labelled edges. By applying this idea, we have seen how to construct paper models of the basic surfaces.

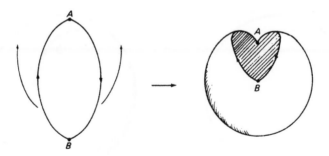

Continuous process of joining 'opposing' points
produces line of *self* -intersection in this
plane projection of *P*

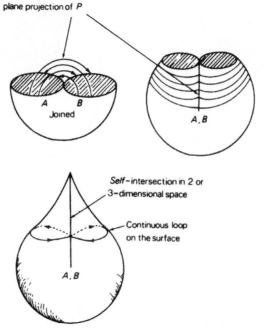

Fig. 2.1.8.

We have given a method of deciding if a compact surface is *non*-orientable by looking at its plane model.

Finally we have stated that it is possible using the connected sum construction to build every compact surface imaginable from the four basic surfaces *S*, *T*, *K* and *P*.

2.6 COMMENTS

It seems reasonable to call the surface *K* a (Klein) bottle, since it could be considered to be a bottle with no inside and no outside!

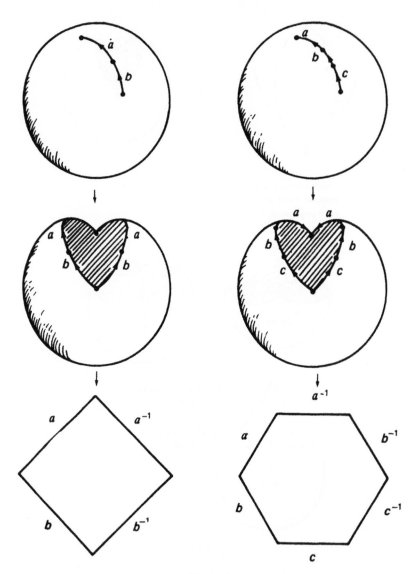

Fig. 2.1.9.

But why should we call the surface P a projective plane? The answer lies in the history of art.

From the 15th century onwards the development by painters of the system of focused perspective led to the growth of a new form of geometry, called **projective geometry**.

The painters were concerned with what we see and with representing this on a flat canvas. Euclidean geometry is inappropriate for this. In Euclidean plane geometry,

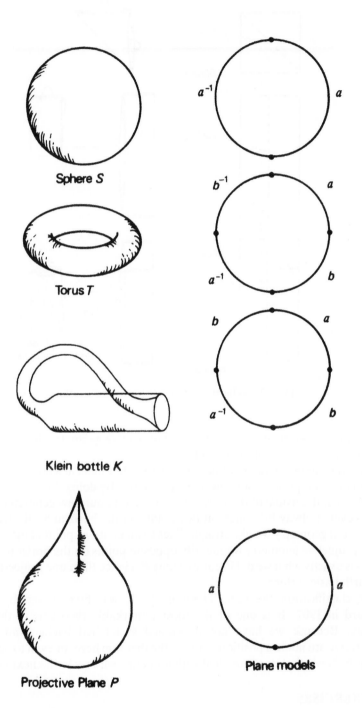

Sphere S

Torus T

Klein bottle K

Projective Plane P

Plane models

Fig. 2.1.10.

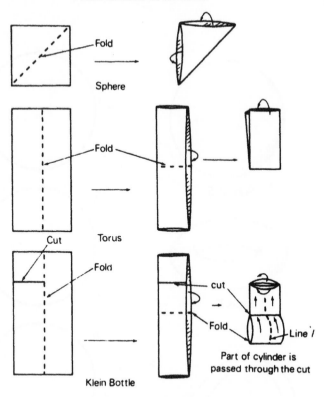

Fig. 2.2.1 — Instructions for constructing basic surfaces.

parallel lines never meet, while in the new projective geometry, in accordance with what we see, every pair of lines meet.

If we stand between railway lines, we see that they meet at a point on the horizon, so we include this point in our new plane geometry. By doing this for parallel lines in *all* directions, the *whole* of the horizon is included in our new geometry.

However, railway lines meet at *two* points on the horizon as illustrated in Fig. 2.6.1. But in geometry we like straight lines to meet at *one* point at most. To ensure this, in projective geometry we identify opposite points on the horizon.

This is exactly what we did in constructing P. Hence the name **projective plane** for our fourth basic surface.

The classification theorem, Theorem 2.4.1, was first proved by Dehn and Heegaard in 1907. It is one of the most remarkable theorems in the whole of topology. Because we have not developed a detailed background of general topology, we are not in a position to prove the theorem here, or even to appreciate its full depth. Nevertheless we do deal with it in our own way in the next chapter.

2.7 EXERCISES

(1) Construct a paper model of the projective plane P using the directions indicated in Fig. 2.2.2(b). Before taping the model, mark on the paper a pair of diametrically

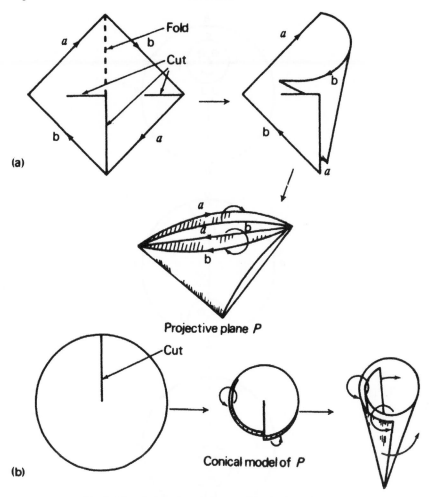

Fig. 2.2.2 — Instructions for constructing basic surfaces.

opposite points p, p' on the circular boundary of the paper. Sketch their semicircular neighbourhoods on the paper, and examine the completed model to see how the identified point p becomes '*locally flat*'.

Sketch a sequence of clockwise arrows on the model, using Fig. 2.3.3 as a guide, to show that P is *non*-orientable.

(2) (a) Describe the loop cuts of Exercise 1.11.3 on a plane model of T.
   (b) Describe the cut of Exercise 1.11.8(c) on a plane model of K.
   (c) On a plane model of K, describe two loop cuts which would produce two interlocking Möbius bands from the associated paper model in 3-dimensional space.

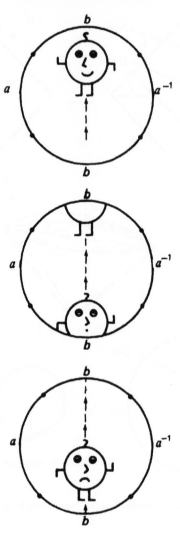

Fig. 2.3.1.

(3) (a) On the plane model of T shown in Fig. 2.7.1, divide the top and bottom edges into n *equal* lengths and mark the points of division $a_0, a_1, \ldots, a_n$. For a given p, $1 \leqslant p \leqslant n$, draw oblique lines on the plane model as follows: link a_0 on the top edge to a_p on the bottom edge; now draw lines parallel to this line so that every point $a_0, a_1, \ldots, a_n$ has such a line passing through it.

        For each value of p, $1 \leqslant p \leqslant n$, determine the number of loops this produces on T.

  (b) Suppose that the plane model of T shown in Fig. 2.7.2 has edges of length 1. For any real number q, $0 \leqslant q \leqslant 1$, describe a path on T as follows. Join the

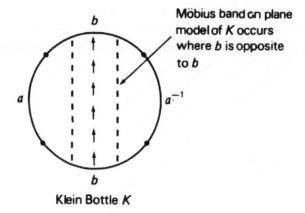

Möbius band on plane
model of K occurs
where b is opposite
to b

Klein Bottle K

Fig. 2.3.2.

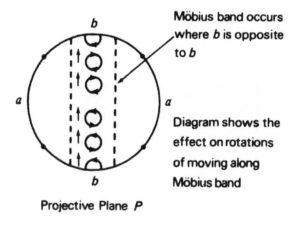

Möbius band occurs
where b is opposite
to b

Diagram shows the
effect on rotations
of moving along
Möbius band

Projective Plane P

Fig. 2.3.3.

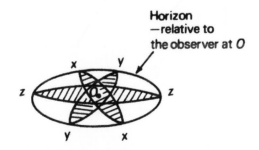

Horizon
—relative to
the observer at O

Fig. 2.6.1 — We show three sets of railway lines in three different directions. In projective
geometry we identify opposing pairs of points.

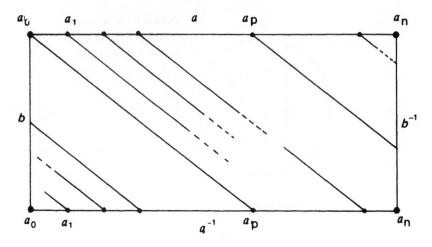

Fig. 2.7.1.

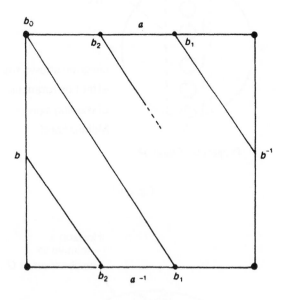

Fig. 2.7.2.

point b_0 on the top edge to the points b_1 distance q from b_0 along the bottom edge. From the point b_1 on the top edge draw a line parallel to this line and suppose that it meets the bottom edge at b_2. Draw a line from b_2 on the top edge parallel to the previous lines. Continue this process. When does this produce a loop on T?

(c) Repeat the procedure described in (b) on a plane model of K. When does this produce a loop on K?

(d) Let $r \geqslant 2$. On a plane model of the projective plane P, choose a point c_0 on the circular boundary and join this, along a chord, to a point c_1 angle $2\pi/r$ clockwise from c_0; see Fig. 2.7.3. Join the diametrically opposite point c_1 to

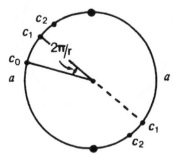

Fig. 2.7.3.

the point c_2, $2\pi/r$ clockwise from c_1, and so on. For which values of r will this produce

  (i) a loop
  (ii) a non-intersecting loop.

(4) (a) Sketch the plane model of $3T$ obtained by slicing the space model shown in Fig. 2.7.4 along the loops a, b, c, d, e, and f.

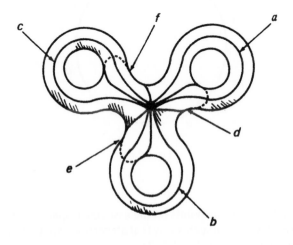

Fig. 2.7.4.

(b) By means of a sequence of sketches show how the sides of the 16-gon in
Fig. 2.7.5 may be identified to form a space model of $4T$. Mark on your

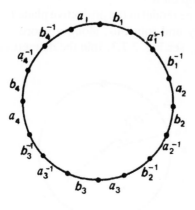

Fig. 2.7.5.

sketch of the space model the *single* point to which the vertices of the 16-gon
correspond.

(5) The vertices of the plane model of a compact surface M shown in Fig. 2.7.6

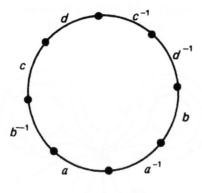

Fig. 2.7.6.

correspond to just *three distinct* points on the associated space model of M. Attach a
label 1, 2, or 3 to each vertex in such a way that all vertices labelled i correspond to the
same point on the space model, $(i = 1, 2, 3)$.

3

Surfaces as plane diagrams

3.1 PLANE MODELS AND THE CONNECTED SUM CONSTRUCTION

We are now on the point of weaning ourselves from our intuitive ideas about surfaces. This will be done by basing our theory on plane models.

To this end let us consider how the connected sum construction is represented in the plane models.

Fig. 3.1.1, which illustrates the construction of $T \# T$ as a plane model, makes this evident.

Let C_1 and C_2 denote the boundaries of the discs on the surfaces T whose interiors are removed in the connected sum construction.

We choose the discs so that C_1 and C_2 pass through points of the surfaces T which are represented by the vertices p_1 and p_2 of the plane models.

This is shown in Fig. 3.1.1(a).

After we have have cut out the interiors of the discs, the resulting sets are represented in the plane by the pentagons of Fig. 3.1.1(b).

Finally, we identify the sides C_1 and C_2 of the pentagons, as shown in Fig. 3.1.1(c).

This process corresponds to the identification of the perimeters of C_1 and C_2 on the space models.

We end up with the octagonal plane model of $2T$ shown in Fig. 3.1.1(d).

Fig. 3.1.2 illustrates the process of obtaining a space model of $2T$ from this plane model.

It is a simple matter to extend the above method to the construction of a plane model of the connected sum of any number of basic surfaces.

In Fig. 3.1.3 we apply the procedure to the surface $T \# K \# P \# T$. (Since the connected sum construction is an associative operation, there is no need to insert brackets.)

We take the component plane models, line them up, join them up, and open out the resulting diagram.

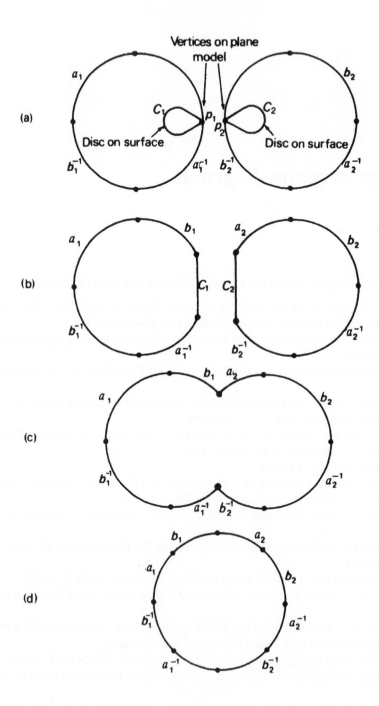

Fig. 3.1.1.

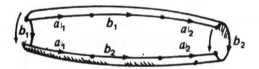

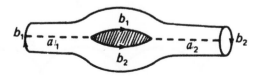

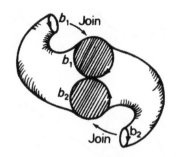

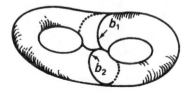

Fig. 3.1.2.

By Theorem 2.4.1, a plane model of any compact surface can be constructed in this way.

3.2 ALGEBRAIC DESCRIPTION OF SURFACES

By listing the labels of the edges of the 14-gon of Fig. 3.1.3(c), in a clockwise sense, we can represent this plane model by the algebraic expression.

$$a_1 b_1 a_2 b_2 a_3 a_4 b_4 a_4^{-1} b_4^{-1} a_3 a_2^{-1} b_2 a_1^{-1} b_1^{-1} \ .$$

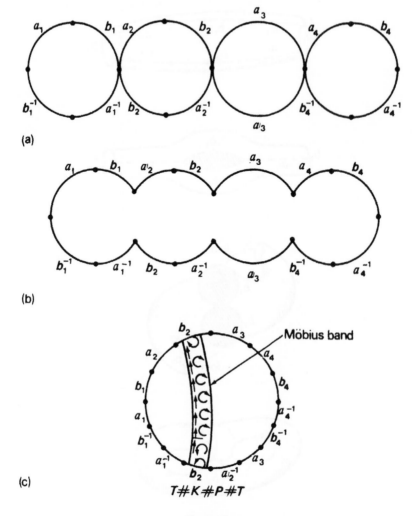

Fig. 3.1.3.

We shall call such an expression a **word** in the constituent symbols a_1, a_2, b_1. etc.

Given a *word*, we can sketch the original plane model, although in doing so we may introduce a harmless rotation.

Thus *words* provide us with another way of representing compact surfaces.

However, such representations are *not unique*, even when we choose the associated polygons to have the least possible number of edges.

For example, $b_1^{-1}a_1^{-1}b_2a_2^{-1}a_3b_4^{-1}a_4^{-1}b_4a_4a_3b_2a_2b_1a_1$, obtained by listing the edges of Fig 3.1.3(c) in anti-clockwise order, also represents

$$T\#K\#P\#T ,$$

and so does $b_1^{-1}a_1b_2a_2a_3a_4b_4a_4^{-1}b_4^{-1}a_3b_2a_2^{-1}b_1a_1^{-1}$.

To obtain the latter, rotate the first two polygons in Fig. 3.1.3(a) through 90° clockwise before constructing the plane model, and then list the labels of the edges in clockwise order.

Fortunately, non-uniqueness is a minor problem. As we shall see, we can live with it quite easily.

In fact, *word* representations of compact surfaces will become a useful tool in the development of the theory.

By section 2.3 and Figs 2.3.2 and 2.3.3, the Möbius band described on the plane model in Fig. 3.1.3(c) shows that $T\#K\#P\#T$ is a *non*-orientable compact surface.

This band links the edges b_2 in the model, which suggests how we might detect *non*-orientability from the *word* representing a compact surface.

A compact surface is *non*-orientable when and only when a *word* representing the surface is of the form: $\ldots\ldots a\ldots\ldots a\ldots\ldots$ for at least one edge a.

We note that each of the *word* representations given above for the surface $T\#K\#P\#T$ is of this form for the edges a_3 and b_2.

DEFINITION 3.2.1 A 2n-gon with edges labelled in pairs $a_i, a_i (i = 1,2,3,\ldots n)$, with possibly the index $^{-1}$ added to some of the labels, is called a **non-orientable** 2n-gon if its *word* representation has the form: $\ldots\ldots a_j\ldots\ldots a_j\ldots\ldots$ for at least one edge a_j, and is called an *orientable* 2n-gon otherwise.

Fig. 3.2.1(a) shows an orientable 10-gon and Fig. 3.2.1(b) shows a *non*-orientable 8-gon.

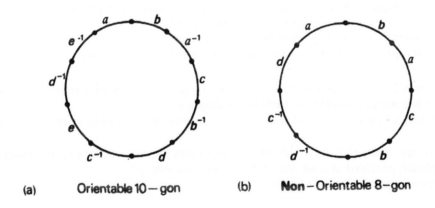

(a)   Orientable 10 – gon   (b)   **Non** – Orientable 8 – gon

Fig. 3.2.1.

3.3   ORIENTABLE 2n-GONS

We know that a plane model of an orientable compact surface is an orientable 2n-gon.

It is natural to ask whether the converse is true, i.e. if we have *any* orientable 2n-gon, is it necessarily the plane model of some orientable compact surface?

It is the purpose of this section to show that the answer is *yes*.

We know from Theorem 2.4.1 that an orientable compact surface is a connected sum of tori. We also know that the *word* representation of the usual plane model of a torus is $aba^{-1}b^{-1}$.

Hence the *word* representation for a connected sum of m tori can be written in the form:

$$a_1 b_1 a_1^{-1} b_1^{-1} a_2 b_2 a_2^{-1} b_2^{-1} \ldots \ldots \ldots a_m b_m a_m^{-1} b_m^{-1}$$

by a suitable extension of Fig. 3.1.1(d) to m tori with appropriate modification of the notation. (This is given as an exercise.)

Our task is done if we can manipulate, in a *reversible* way, the *word* representation of an orientable 2n-gon, *without affecting any surface represented*, until it looks like the above normal *word* representation of mT, for some value of m.

The operations by which we perform this manipulation must be reversible and must be such that the *word* reached at *each* stage of the manipulation represents the *same* surface, up to homeomorphism.

The following operations satisfy these critreria.

Operation 1.   The cyclic permutation of the edges in a *word*

This can be expressed by the **identity 1** $AB \equiv BA$ where A and B denote strings of edges.

For example, $a_1 b_1 a_1^{-1} b_1^{-1}$ and $b_1^{-1} a_1 b_1 a_1^{-1}$ represent the same surface.

In fact they represent the same plane model up to rotation.

In general, any cyclic permutation of a *word* amounts to a rotation of the plane model it represents and hence does not alter the surface represented.

Operation 2.   The removal or introduction of a pair of consecutive edges xx^{-1} (or $x^{-1}x$) in a *word*, provided the *word* is not simply xx^{-1}.

In the following, a capital letter denotes a sequence (possibly empty) of edge labels.

We are saying that $Axx^{-1}B$ and AB represent the *same* surface up to homeomorphism, provided at least one of A and B is non-empty.

We express this by the **identity 2**

$$Axx^{-1}B \equiv AB \ .$$

This is justified in Fig. 3.3.1. It really amounts to saying that $M\#S$ is homeomorphic to M, for *any* surface M.

Operation 3.   The replacement of $AxBCx^{-1}$ by $AyCBy^{-1}$ and vice versa, where we write $y = xB$ to mean that the edge y connects the same pair of vertices as does the sequence of edges xB and $x^{-1} = By^{-1}$ to mean that the edge x^{-1} connects the same

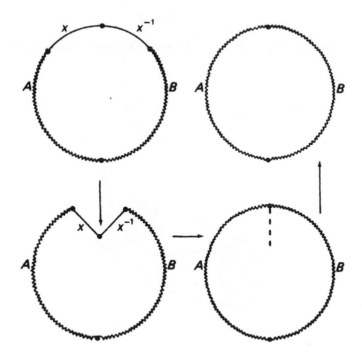

Fig. 3.3.1.

pair of vertices as the sequence By^{-1}, *due regard being paid to direction as indicated by the indices.*

The above operation may be expressed by the **identity 3**

$$AxBCx^{-1} \equiv AyCBy^{-1} .$$

This is justified in Fig.3.3.2, where we see that the first stage in constructing a space model of the surface is the same, as shown in Fig. 3.3.2(b), whether we begin by identifying the edges x of Fig. 3.3.2(a) or the edges y of Fig. 3.3.2(c).

All three operations are clearly reversible.

LEMMA 3.3.1   Every orientable 2n-gon is a plane model of some orientable compact surface.

PROOF
Let M_0 denote the *word* representation of an orientable 2n-gon.

Let us say that the edges x,x^{-1} and y,y^{-1} in M_0 are **separated pairs** if $M_0 = AxByCx^{-1}Dy^{-1}$, possibly after a cyclic permutation of the edges.

If separated pairs do exist, then M_0 contains a *word* representation of the torus which we can isolate by repeated use of Operations 1 and 3 as follows:

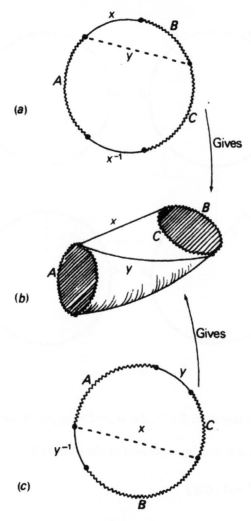

Fig. 3.3.2.

$$M_0 \equiv Ax_1yCBx_1^{-1}Dy^{-1} \quad \text{(Op 3)}$$
$$\equiv Ax_1y_1x_1^{-1}DCBy_1^{-1} \quad \text{(Op 3)}$$
$$\equiv y_1x_1^{-1}DCBy_1^{-1}Ax_1 \quad \text{(Op 1)}$$
$$\equiv y_1x_2^{-1}y_1^{-1}ADCBx_2 \quad \text{(Op 3)}$$
$$\equiv x_2y_1x_2^{-1}y_1^{-1}ADCB \quad \text{(Op 1)}$$

Now put $x_2y_1x_2^{-1}y_1^{-1} = a_1b_1a_1^{-1}b_1^{-1}$ and $M_1 = ADCB$.

If M_1 contains separated pairs we repeat the procedure to get:

$$M_1 = a_2b_2a_2^{-1}b_2^{-1}M_2 \; ,$$

We continue in this way until we reach a form:

$$M_{m-1} = (a_m b_m a_m^{-1} b_m^{-1}) M_m \ ,$$

where M_m contains *no* separated pairs.

Let us turn our attention to M_m.

If M_m is empty, we are finished. Otherwise we argue as follows.

Since M_0 is the *word* representation of an *orientable* 2n-gon, we can write:

$$M_m = p_1 V_1 p_1^{-1} U_1 \quad \text{for some } p_1 \ ,$$

where V_1 has *no* separated pairs.

Similarly we can write:

$$V_1 = p_2 V_2 p_2^{-1} U_2 \ ,$$

where V_2 has no separated pairs.

Continuing in this way, we eventually reach an integer k with $V_k = z z^{-1}$.

If $M_m \neq V_k$, then Operation 2 enables us to remove V_k.

By repeating this procedure we eventually reduce M_m to the form $w w^{-1}$.

If $m > 0$, then Operation 2 enables us to remove $w w^{-1}$, to obtain:

$$M_0 = a_1 b_1 a_1^{-1} b_1^{-1} \ldots\ldots\ldots a_m b_m a_m^{-1} b_m^{-1} \ ,$$

which is the *word* representation of an m-holed torus.

If $m = 0$, then $M_0 = w w^{-1}$, which is the *word* representation of a sphere.

We have thus converted M_0 into the *word* representation of either the surface mT or the surface S by the use of Operations 1, 2, and 3, each of which is reversible.

By carrying out the sequence of operations in reverse order, we can convert our final *word* representation of a surface into the original *word* representation M_0 of an orientable 2n-gon.

Since the operations do *not* alter, *up to homeomorphism*, the surface represented, it follows that M_0 is the *word* representation of an orientable compact surface.

This proves the lemma.

3.4 NON-ORIENTABLE 2n-GONS

Similar results hold for *non*-orientable 2n-gons. More precisely we have:

LEMMA 3.4.1 Every *non*-orientable 2n-gon is a plane model of some *non*-orientable compact surface.

To prove this result we require one more operation.

Operation 4.   The replacement of $AxBxC$ by $AyyB^{-1}C$ and vice versa, where $y = xB$ and $x = yB^{-1}$.

We express this operation by **identity 4**

$$A\,x\,B\,x\,C \equiv A\,y\,y\,B^{-1}\,C\;.$$

Here B^{-1} represents the sequence of edges B read in the *reverse order* and with the *opposite direction* imposed on *each* of them.

For example, if $B = ab^{-1}c$, then $B^{-1} = c^{-1}ba^{-1}$.

This operation is justified in Fig. 3.4.1. The first stage of forming a space model of the surface as shown in Fig. 3.4.1(b) is the same whether we begin by identifying the edges x of Fig. 3.4.1(a) or the edges y of Fig. 3.4.1(c).

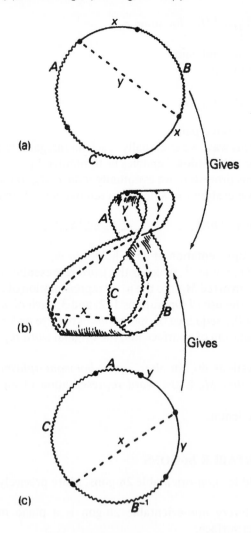

Fig. 3.4.1.

This may be difficult to see from the plane figure. Non-believers are invited to construct the Möbius band of Fig. 3.4.1(b), and then to make the cut necessary to produce the polygon of Fig. 3.4.1(c).

This new operation is clearly reversible.

To prove the Lemma, suppose M is a *word* representation of a non-orientable 2n-gon. By applying Operation 4 and possibly Operation 1 we can write $M \equiv a_1 a_1 M_1$ for some *word* M_1 and edge a_1. If M_1 is non-orientable then $M_1 \equiv a_2 a_2 M_2$ say, and so on. Eventually we reach a *word* M_m which is either empty or orientable. If M_m is empty then M has been expressed as the connected sum of m projective planes. If M_m is non-empty then by Lemma 3.3.1 it can be expressed either as a sphere or as a connected sum of tori. Then M can be expressed as the connected sum of m projective planes and this orientable surface.

3.5   THE WORKING DEFINITION OF A SURFACE

By assuming Theorem 2.4.1, as yet unproved, we have a fairly complete understanding of compact surfaces in terms of our intuitive definitions and 2n-gons with labelled edges.

The polygons enable us to make precise our intuitive ideas and are potentially useful in solving problems about surfaces.

In this section, we use them to obtain a more practical mathematical model of the compact surfaces in which we are interested.

With this in mind, we give the following definition of a *combinatorial surface*.

DEFINITION 3.5.1   **An orientable compact surface** is an orientable 2n-gon.

A non-orientable compact surface is a *non*-orientable 2n-gon.

In future we use symbols like M to denote *both* a compact surface *and* its *word* representation.

In the light of our new definition of a compact surface, we redefine several concepts met before.

DEFINITION 3.5.2   The compact surfaces M_1 and M_2 are **homeomorphic**, written $M_1 \equiv M_2$, if M_1 can be converted into the form M_2 by the Operations 1 to 4 of sections 3.3 and 3.4.

DEFINITION 3.5.3   The **connected sum** of the compact surfaces M_1 and M_2 is the surface $M_1 M_2$.

DEFINITION 3.5.4   We call

aa^{-1}	a **sphere**	denoted by S ,
$aba^{-1}b^{-1}$	a **torus**	denoted by T ,
aa	a **projective plane**	denoted by P ,
$aba^{-1}b$	a **Klein bottle**	denoted by K .

The compact surfaces, S, T, P, and K are called the **basic surfaces**.

$nM = MMMM\ldots\ldots M$ denotes the connected sum of n *copies* of the single basic surface M.

By convention, $0T$ is identified with S.

DEFINITION 3.5.5   A compact surface M with the appropriate edges identified in $\mathbb{R}^n(n \geqslant 3$ if the surface is orientable and $n \geqslant 4$ if it is *non*-orientable) is called a **space model** of M. *Note*: $\mathbb{R}^n$ denotes the usual n-dimensional real Euclidean space i.e. a generalization to n dimensions of the 'ordinary' 3-dimensional Euclidean space of everyday life.

3.6   THE CLASSIFICATION THEOREM

Let us consider again our fundamental theorem in the light of our new definitions.

THEOREM 3.6.1   *The Classification Theorem.*
- (a)   An orientable compact surface is homeomorphic to nT for some $n \geqslant 0$.
- (b)   A *non*-orientable compact surface is homeomorphic to either $(nT)K$ or $(nT)P$, for some $n \geqslant 0$.

PROOF
- (a)   We proved (a) when we proved Lemma 3.3.1.
- (b)   In proving Lemma 3.4.1, we saw that *any non*-orientable compact surface M is homeomorphic to a surface of form:

$$(nT)\ (mP)$$

for some integers $n \geqslant 0$ and $m \geqslant 1$.

The proof now proceeds by induction on m.

If $m = 1$, then M takes one of the forms in (b) for *any* value of n.

By the induction hypothesis, *either* $(nT)\ (kP) \equiv (qT)P$, or $(nT)\ (kP) \equiv (qT)K$, for *some* q, where $k \geqslant 1$.

Suppose $(nT)(kP) \equiv (qT)P$.
Then $\begin{aligned}[t]
(nT)((k+1)P) &\equiv (qT)(2P) \\
&\equiv (qT)(aabb) \\
&\equiv (qT)(a_1 b^{-1} a_1 b) \quad \text{(Op 4)} \\
&\equiv (qT)K \qquad\qquad \text{(Op 1)}
\end{aligned}$

On the other hand, if $(nT)(kP) \equiv (qT)K$ for *some* q, then

$$\begin{aligned}
(nT)((k+1)P) &\equiv (qT)(abab^{-1}cc) \\
&\equiv (qT)(a_1 a_1 b^{-1} b^{-1} cc) & \text{(Op 4)} \\
&\equiv (qT)(a_1 a_1 b_1 c^{-1} b_1 c) & \text{(Op 4)} \\
&\equiv (qT)(a_2 b_1^{-1} a_2 c^{-1} b_1 c) & \text{(Op 4)} \\
&\equiv (qT)(a_2 b_2^{-1} c^{-1} a_2 b_2 c) & \text{(Op 3)} \\
&\equiv (qT)(a_2 b_2^{-1} c_1^{-1} b_2 a_2 c_1) & \text{(Op 3)}
\end{aligned}$$

$$\equiv (c_1^{-1}b_2a_2c_1)(qT)(a_2b_2^{-1}) \qquad \text{(Op 1)}$$
$$\equiv (c_1^{-1}b_3)(qT)(a_2a_2c_1b_3^{-1}) \qquad \text{(Op 3)}$$
$$\equiv (c_1b_3^{-1}c_1^{-1}b_3)(qT)(a_2a_2) \qquad \text{(Op 1)}$$
$$\equiv ((q+1)T)P \qquad \text{(Op 1)}$$

Hence, if, for any value of n, $(nT)(kP)$ can be expressed in one of the forms in (b), then so can $(nT)((k+1)P)$ for any value of n.

An appeal to induction completes the proof.

DEFINITION 3.6.1   A compact surface, expressed in one of the forms given in Theorem 3.6.1, is said to be in *normal* form.

Note that in the course of the above proof, we have shown that $2P \equiv K$ and $KP \equiv TP$.

These facts, together with the *commutativity* and *associativity* of the connected sum operation, are useful in converting an awkward expression for a compact surface into the *normal* form.

EXAMPLE 3.6.1   Express $TPKP(2K)T$ in *normal* form.

Solution
$$\begin{aligned}
TPKP(2K)T &\equiv T(PK)(PK)KT \\
&\equiv T(TP)(TP)KT \\
&\equiv (4T)P(PK) \\
&\equiv (4T)P(TP) \\
&\equiv (5T)(2P) \\
&\equiv (5T)K.
\end{aligned}$$

3.7 SUMMARY

By interpreting the connected sum construction in terms of plane models, we were able to show how to form a plane model of *any* connected sum of basic surfaces.

Assuming the classification theorem, this enabled us to form a plane model of *any* compact surface.

Bearing in mind this largely intuitive background, and the connections with 2n-gons, we then laid a firmer foundation for our theory by *defining* a compact surface for our purposes to *be* a 2n-gon.

Finally, in the new theory, we stated and proved *our version* of the classification theorem.

3.8 EXERCISES

(1) Forming the connected sum of four tori along the lines of the construction shown in Fig. 3.1.3 would produce a *word* expression

$$a_1b_1a_2b_2a_3b_3a_4b_4a_4^{-1}b_4^{-1}a_3^{-1}b_3^{-1}a_2^{-1}b_2^{-1}a_1^{-1}b_1^{-1}$$

for $4T$.

Illustrate the sequence of connected sums you would form to produce the word

$$a_1 b_1 a_1^{-1} b_1^{-1} a_2 b_2 a_2^{-1} b_2^{-1} a_3 b_3 a_3^{-1} b_3^{-1} a_4 b_4 a_4^{-1} b_4^{-1}$$

representing $4T$ (i.e. the *normal* form for $4T$).

(2) List all possible *words* in the symbols a_1, a_2, b_1, b_2 which represent $2T$.

(3) Sketch plane models and write down *words* representing
    (a) $2T2K$ (b) $3T2P$ (c) $3K2P$.

(4) Show that if the compact surface M is *not* a sphere then in the normal form for M the vertices of the plane model correspond to just *one* point on the space model of M.

(5) Fig. 3.8.1 shows how we can justify Operation 3 by means of a sequence of

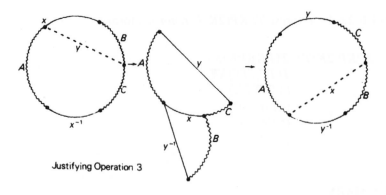

Justifying Operation 3

Fig. 3.8.1.

manipulations of the plane model. In a similar way
    (a) justify Operation 4
    (b) show $2P \equiv K$.

(6) (a) Using Operations 1 to 4, prove
      (i) $2P \equiv K$ and   (ii) $KP \equiv TP$.
    (b) Use these equivalences to express the surfaces in question (3), above, in normal form.

(7) Prove that any *non*-orientable compact surface M can be expressed in the form $M \equiv nP$, for some $n \geqslant 1$.

(8) Which of the following are compact surfaces?

(a) $b^1 a_2 b_3 a_3^{-1} b_3^{-1} a_3 a_2^{-1} a_1^{-1} b_1^{-1} a_1$

(b) $a_1 a_2^{-1} a_1^{-1} a_3 a_4 a_2 a_4^{-1}$

(c) $a_1 a_2 a_3^{-1} a_4 a_5^{-1} a_6 a_7 a_7^{-1} a_6^{-1} a_5^{-1} a_4^{-1} a_3^{-1} a_2^{-1} a_1^{-1}$

(d) $a_1 a_2 . a_3^{-1} a_4 a_5 a_2^{-1} a_1^{-1} a_4^{-1} a_3 a_5$

(e) $a_1 a_2 a_1 a_3$

(f) $a_1 a_2 a_3 a_2 a_4 a_5^{-1} a_4 a_5 a_3^{-1} a_1 a_6 a_7 a_6 a_7^{-1}$

For those words which do *not* represent compact surfaces, how would you recognize this fact from the associated space model? For those words which *do* represent compact surfaces, state whether the surface is orientable or non-orientable, and using Operations 1 to 4 express it in its normal form.

(9) Give instructions for forming paper models of

   (a) *TP* from a hexagonal sheet of paper,
   (b) *TK* from an octagonal sheet of paper.

(10) How many *T*'s in TOPOLOGY?

4

Distinguishing surfaces

4.1 INTRODUCING $\chi(M)$

Two compact surfaces M_1 and M_2 are homeomorphic if and only if the *word* M_1 can be converted into the *word* M_2 by using the Operations 1 to 4 of sections 3.3 and 3.4.

In practice, this conversion may not be easy to carry out. It is, therefore, convenient to have a simple method which enables us to determine when two surfaces are homeomorphic. In this chapter, we associate with each compact surface an integer which helps to provide just such a method.

Given the compact surface M, that is a 2n-gon for some n, we count the number v of points on the space model of M represented by the 2n vertices of the plane model.

DEFINITION 4.1.1   The **Euler characteristic** of the compact surface M is the integer

$$\chi(M) = v - n + 1 .$$

To illustrate this, let us consider the compact surface $M = abcdc^{-1}ba^{-1}ed^{-1}e^{-1}$. The 10-gon with directed edges identified in pairs, shown in Fig. 4.1.1, represents this compact surface.

First a definition:

DEFINITION 4.1.2   In Fig. 4.1.1, we call the vertex p_1 the **initial** vertex and p_2 the **final** vertex of the edge a.

Now let us count the number of points on the space model of M that are represented by the vertices of the plane model shown in Fig. 4.1.1. We do this as follows.

First attach the label 1 to the final vertex of a at the top of the figure. When the edges a are identified, this vertex is attached to the final vertex of the second edge a at

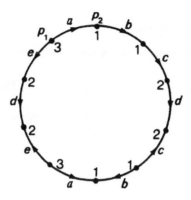

Fig. 4.1.1.

the bottom of the figure. Hence we attach the same label 1 to this second vertex. *Notice* that this same vertex is the final vertex for the edge *b*. Thus we label the final vertex of the second edge *b* with 1. This is the initial vertex of the edge *c* near the top of the figure, so we next label the initial vertex of the second *c* with 1. The latter is also the initial vertex of the edge *b* near the bottom of the figure. Since the initial vertex of the other edge *b* has already been labelled with 1, the process terminates.

The four vertices labelled 1 represent a *single* point on the space model of the surface *M*.

Next we choose an unlabelled vertex, label it 2, and repeat the above procedure. Continuing in this way, we eventually arrive at the labelling shown in Fig. 4.1.1. This shows that the vertices of the plane model represent 3 points on the space model of *M*.

Hence, in this case $\chi(M) = 3 - 5 + 1 = -1$.

The Euler characteristic contains vital information about the surface. It turns up again and again in our applications.

In Fig. 4.1.2, we repeat the above procedure to determine the Euler characteristic of each of the basic surfaces. We have:

$$\chi(S) = 2$$
$$\chi(T) = 0$$
$$\chi(K) = 0$$
$$\chi(P) = 1$$

Notice that although homeomorphic surfaces have the same Euler characteristic *the converse is false*, as shown by $\chi(T) = \chi(K)$ but $T \not\equiv K$.

4.2   χ(*M*) AND THE CONNECTED SUM CONSTRUCTION

Since the connected sum construction is central to the classification theorem, we expect its relation with the Euler characteristic to be of some importance in our theory. This relation is the content of

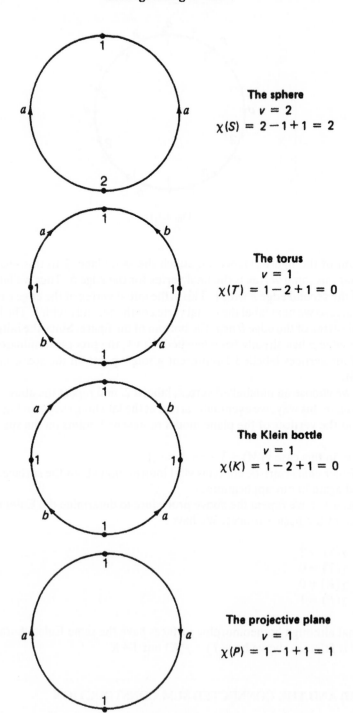

Fig. 4.1.2.

THEOREM 4.2.1 For any compact surfaces M_1 and M_2

$$\chi(M_1 M_2) = \chi(M_1) + \chi(M_2) - 2 .$$

PROOF Suppose that v_i points on the space model of the surface M_i are represented by the vertices of the $2n_i$-gon, where $i = 1$ or 2.

As we can see in Fig. 4.2.1, when we form $M_1 M_2$, one of the vertices of M_1 is identified with one of the vertices of M_2. Thus the vertices of the constructed $2(n_1 + n_2)$-gon represent $v_1 + v_2 - 1$ points on the space model of $M_1 M_2$.

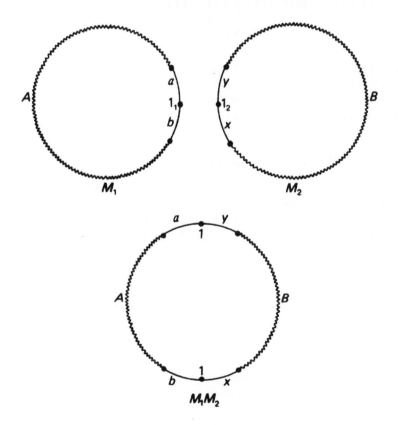

Fig. 4.2.1.

Hence

$$\begin{aligned}
\chi(M_1 M_2) &= (v_1 + v_2 - 1) - (n_1 + n_2) + 1 \\
&= (v_1 - n_1 + 1) + (v_2 - n_2 + 1) - 2 \\
&= \chi(M_1) + \chi(M_2) - 2 .
\end{aligned}$$

COROLLARY 4.2.1

> (a) $\chi(nT) = 2 - 2n$
> (b) $\chi((nT)K) = -2n$
> (c) $\chi((nT)P) = 1 - 2n$.

PROOF   (a) The proof is by induction on n. The result is certainly true for n = 1, so let us suppose it is true for n = k.

Using Theorem 4.2.1, we have: ·

$$\chi((k + 1)T) = \chi((kT)T) = \chi(kT) + \chi(T) - 2$$
$$= 2 - 2k + 0 - 2$$
$$= 2 - 2(k + 1) ,$$

which is the required result for n = k + 1.

An appeal to induction completes the proof.

(b) and (c)   Since we know now the values of $\chi(nT)$, $\chi(P)$ and $\chi(K)$, a direct application of Theorem 4.2.1 proves (b) and (c).

4.3   HOW TO TELL THE DIFFERENCE

Now we are in a position to determine quickly and easily precisely when two compact surfaces are homeomorphic.

As noted at the end of section 4.1, although homeomorphic surfaces have the same Euler characteristic, the converse is false. However, a surprisingly small addition to the Euler characteristic suffices to give necessary and sufficient conditions for two compact surfaces to be homeomorphic. Formally we have:

THEOREM 4.3.1   The compact surfaces M_1 and M_2 are homeomorphic if and only if

(a) $\chi(M_1) = \chi(M_2)$ *and*
(b) *either both* M_1 *and* M_2 *are orientable.*
    or both M_1 *and* M_2 *are* non-*orientable.*

PROOF   Suppose M_1 and M_2 are homeomorphic. Then by Definition 3.5.2, we can convert M_1 into M_2 by application of Operations 1 to 4 of sections 3.3 and 3.4.

To prove (a) and (b) above, we show that Operations 1 to 4 *do not affect* either the orientability or the Euler characteristic of the compact surface to which they are applied.

It is clear that Operations 1 to 4 can neither introduce nor remove a pair of edges *a* and *a*. Thus an orientable surface remains orientable and a *non*-orientable surface remains *non*-orientable after the application of any of the Operations 1 to 4.

Now let us consider the effect of these operations on the Euler characteristic.

Operation 1 does not change the number of edges of the polygon on which it acts; nor does it change the number of points on the space model of the surface represented by the vertices of the polygon.

Hence Operation 1 leaves the Euler characteristic of the compact surface unchanged.

By examining Fig. 3.3.1, we see that under Operation 2 we lose a point on the space model represented by a single vertex. However, we also lose a single pair of edges. Thus the Euler characteristic of the compact surface remains the same.

Examination of Figs 3.3.2 and 3.4.1 shows that under Operations 3 and 4, the number of points on the space model represented by the vertices of the polygon remains unchanged, while one pair of edges is removed and a new pair of edges is introduced. Once again, on balance, the Euler characteristic remains the same.

Altogether, we have proved that if $M_1 \equiv M_2$, then $\chi(M_1) = \chi(M_2)$ and *either* M_1 and M_2 are orientable, *or* M_1 and M_2 are *non*-orientable.

To prove the converse we appeal to Theorem 3.6.1. Suppose that M_1 and M_2 are orientable and $\chi(M_1) = \chi(M_2)$. By Theorem 3.6.1, $M_1 \equiv mT$ and $M_2 \equiv nT$ for some non-negative integers m and n. By Corollary 4.2.1, $\chi(M_1) = 2 - 2m$ and $\chi(M_2) = 2 - 2n$. But we are given that $\chi(M_1) = \chi(M_2)$. Thus $2 - 2m = 2 - 2n$. Hence n = m. Thus $M_1 \equiv nT \equiv M_2$, and M_1 and M_2 are homeomorphic.

Finally, suppose M_1 and M_2 are *non*-orientable and $\chi(M_1) = \chi(M_2)$. By Theorem 3.6.1, $M_1 \equiv mTK$ or mTP and $M_2 \equiv nTK$ or nTP for some non-negative integers m and n. By Corollary 4.2.1, $\chi(M_1) = -2m$ or $1 - 2m$ and $\chi(M_2) = -2n$ or $1 - 2n$. Moreover $\chi(M_1) = \chi(M_2)$. Thus *either* $-2m = -2n$ *or* $1 - 2m = 1 - 2n$. (Other possibilities like $-2m = 1 - 2n$ are ruled out because $-2m$ is *even* and $1 - 2n$ is *odd*.)

Hence we always have n = m. But then either $M_1 \equiv mTK \equiv M_2$ or $M_1 \equiv mTP \equiv M_2$. Hence always $M_1 \equiv M_2$, as required.

This completes the proof of the theorem.

4.4   CAN YOU TELL THE DIFFERENCE?

By evaluating the Euler characteristic and considering orientability, we can easily recognize any compact surface expressed as a plane model. But it is not always easy to identify a surface when presented as a space model. We know nT has n 'holes', but sometimes it is difficult to tell the number of 'holes' a space model has.

Recognizing a space model of a sphere causes no problems even if the space model is as strange as Alexander's horned sphere shown in Fig. 4.4.1. Similarly there is no problem in identifying a space model of T or $2T$ in 3-dimensional space. But how many 'holes' have the space models shown in Fig. 4.4.2?

A simple change of perspective, as given in Fig. 4.4.3, shows us that Fig. 4.4.2(a) is a space model of $3T$. Unfortunately, changing our viewpoint does not help us to recognize the space model in Fig. 4.4.2(b).

To identify this space model, we invoke the connected sum construction. As shown in Fig. 4.4.4, connecting T to *any* compact surface M may be viewed as adding a 'handle' to a space model of M. Thus, by removing handles from a space model of

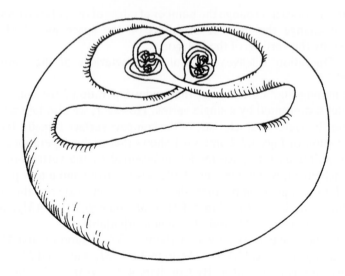

Alexander's horned sphere

Fig. 4.4.1.

an orientable compact surface, we can count the number of tori that collectively form the surface.

In this way, as shown in Fig. 4.4.5, we can deduce that Fig. 4.4.2(b) is in fact a space model of $6T$. It has 6 'holes'.

The number of handles we must connect to a space model of a sphere to obtain a space model of the orientable compact surface M is called the **genus** of M. Thus if $M \equiv nT$, then genus of M is n.

As above, we can find the genus of any orientable compact surface when the latter is realized as a space model in $\mathbb{R}^3$.

4.5   COMMENTS

The formula $V - E + F = 2$ for a closed convex polyhedron, where V is the number of vertices, E is the number of edges, and F is the number of faces, was known to Déscartes in 1639. But the result was first published by Euler one hundred years later. For this reason $\chi(M) = V - E + F$, in the usual notation, is called the Euler characteristic.

Riemann's work on complex function theory played a large part in the growth of topology. In this work, he introduced the idea of a Riemann surface, and related the equivalence of orientable compact surfaces to the genus of the surface. The latter is related to the Euler characteristic and is basically the number of 'holes' of the surface, e.g. a torus has one 'hole'.

Thus Riemann in 1857 could be said to have produced a large part of Theorem 4.3.1(a), bearing in mind that two orientable compact surfaces have the same genus if and only if they have the same Euler characteristic.

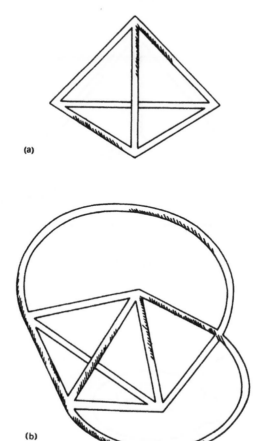

(a)

(b)

Fig. 4.4.2.

In 1861, Möbius published a result close to this theorem. In fact, many mathematicians over the years contributed to the full story of the topological equivalence of compact surfaces.

Having discovered that the sphere is orientable but the projective plane is not, Klein clarified the situation by publishing in 1874 the full result that two *orientable* compact surfaces are homeomorphic if and only if they have the same genus. Our Theorem 4.3.1(a) is this result expressed in terms of the Euler characteristic.

It is still unknown whether a similar approach will work for *all* higher dimensional 'surfaces', i.e. spaces that *locally 'resemble'* $\mathbb{R}^n$, or more precisely are *locally homeomorphic* to $\mathbb{R}^n$. *Note* that '*being locally homeomorphic to* $\mathbb{R}^2$' is just '*being locally flat*' as discussed in Chapter 1.

4.6 EXERCISES

(1) For each of the samples given in Exercise 3.8.3 count the number of points on the surface represented by the vertices of the plane model. Hence write down the Euler

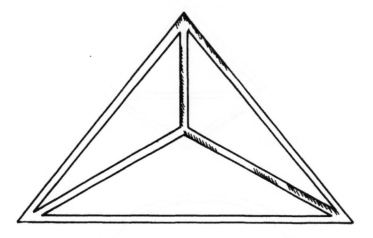

Fig. 4.4.3.

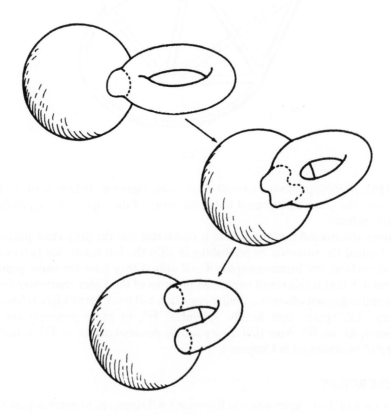

Fig. 4.4.4.

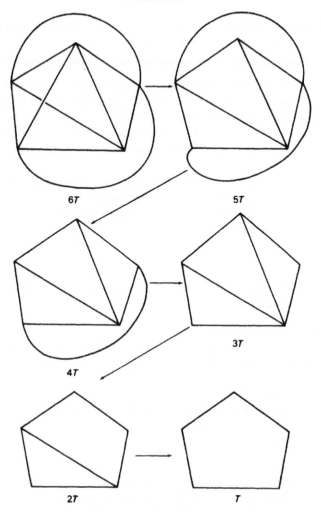

6T   5T

4T   3T

2T   T

Fig. 4.4.5 — Line diagram of Fig. 4.4.2(b).

characteristic of each surface and, using Corollary 4.2.1, check the results you obtained in Exercise 3.8.6(b).

(2) Determine the Euler chracteristic of each of the following compact surfaces and hence express each as a connected sum of *basic* surfaces.

(a) $a_1a_2a_3a_4a_2a_5a_1a_6a_7a_4^{-1}a_7^{-1}a_8a_3a_9a_6a_5^{-1}a_9a_8^{-1}$

(b) $a_1a_2a_3a_2^{-1}a_4a_5a_1^{-1}a_6a_4^{-1}a_3^{-1}a_7a_8a_6^{-1}a_9a_5^{-1}a_8^{-1}\ a_7^{-1}a_9^{-1}$

(c) $a_1a_2a_3^{-1}a_4^{-1}a_5a_6^{-1}a_6a_5^{-1}a_4a_3a_2^{-1}a_1^{-1}$

(d) $a_1a_2a_3a_4a_5a_6a_1a_2^{-1}a_3a_7a_8^{-1}a_9a_{10}a_{11}a_{12}^{-1}a_7a_8a_9^{-1}a_4a_5a_6a_{10}a_{11}a_{12} \rightarrow$

$\rightarrow a_{13}a_{14}^{-1}a_{15}a_{13}a_{16}a_{17}a_{14}a_{18}^{-1}a_{15}a_{18}a_{17}a_{16}$

(e) *minnim.*

(3) Repeat Exercise (2) for the examples of compact surfaces given in Exercise 3.8.8 and thus check the results you obtained in Exercise 3.8.8.

(4) (a) If $kT\, mP \equiv nTK$ show that m is *even* and express n in terms of k and m.
   (b) The surface M can be expressed as a connected sum of projective planes, Klein bottles, and tori in such a way that
      (i) the number of projective planes in the expression is equal to one third of the number in an expression for M in terms of projective planes alone,
      (ii) the number of Klein bottles in the expression is half of the number in an expression for M involving Klein bottles only,
      (iii) the number of tori in the expression is one fifth of the number of tori in the *normal* form for M.
      Find M.

(5) Write down the equation relating the genus and the Euler characteristic of an orientable compact surface.

(6) (a) Call a wire cube A_1. Call a wire lattice of the form shown in Fig. 4.6.1(a), built up from 2^3 wire cubes, A_2. Call a wire lattice built up in this way from n^3 cubes: A_n.
      Find the genus of A_n.
   (b) B_n denotes the surface of the wire framework shown in Fig. 4.6.1(b), with $2n + 1$ linked wire cubes along the base.
      Find the genus of B_n.

(7) (a) Let D_n denote the surface of a wire model of an n-dimensional triangle. Fig. 4.6.2(a) shows D_2, D_3, and D_4.   Find the genus of D_n.
   (b) Let C_n denote the surface of a wire model of an n-dimensional cube. Fig. 4.6.2(b) shows C_4.
      Find the genus of C_n.

(8) (a) Using the labels $a_1, a_2, \ldots\ldots, a_n$ for n of the edges of a 2n-gon,
      (i) how many orientable 2n-gons are possible,
      (ii) how many *non*-orientable 2n-gons are possible?
   (b) (i) How many distinct (non-homeomorphic) orientable compact surfaces can be represented by a 2n-gon?
      (ii) How many distinct *non*-orientable compact surfaces can be represented by a 2n-gon?

(9) Use Theorem 4.3.1 to show $2P \equiv K$; $2K = TK$.

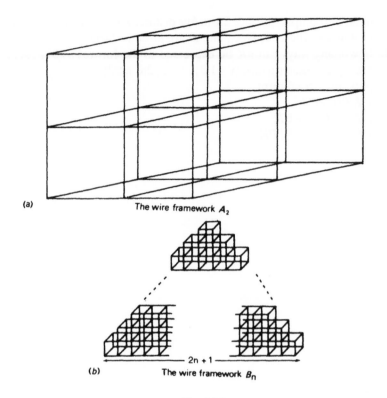

(a)   The wire framework A_2

                    $2n + 1$
(b)   The wire framework B_n

Fig. 4.6.1.

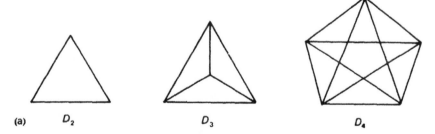

(a) D_2 D_3 D_4

Diagrams of n–dimensional triangles D_n; n = 2,3,4

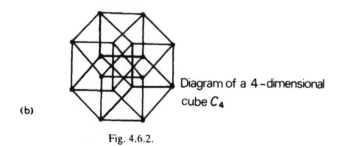

(b)

Diagram of a 4–dimensional cube C_4

Fig. 4.6.2.

(10) (a) Express the compact surface $2K$ in *five* distinct ways (ignoring order) in the form $ABCD$, where A, B, C, and D are *basic* surfaces.

(b) In how many ways can nK be expressed in the form $A_1A_2A_3 \ldots . . A_{2n}$, where A_i is a basic surface ($i = 1,2, \ldots . . , 2n$; $n \geqslant 2$)?

(c) Express in normal form
   (i) $KT2PS3TK$.
   (ii) $K2T3P4S5K6T$.
   (iii) $STPKSTPKSTPK$.

5

Patterns on surfaces

5.1 PATTERNS AND χ(*M*)

The kind of pattern that can be drawn on a surface is to a large extent determined by the nature of the surface. In this chapter we spend some time examining the relation between pattern and surface. But first let us say precisely what we mean by a pattern on a surface. We do this by way of the following definitions.

DEFINITION 5.1.1   A **polygon** is a set of points topologically equivalent to a plane disc with at least two vertices on the boundary. The arcs between the vertices on the boundary are called the **edges** of the polygon.

DEFINITION 5.1.2   A **pattern** on a compact surface *M* is a mosaic of polygons *covering* the *space model* of *M* in such a way that each polygon is contractible to a point on the surface and

P1: two distinct polygons meet at vertices, along *complete* edges, or not at all;
P2: the polygons do not meet themselves.

In Fig. 5.1.1, the polygons in (a), (b), (c), and (d) can *not* form part of a pattern on a compact surface. Condition P1 is violated in (a) and (b), while condition P2 is violated in (c) and (d). However, (e) and (f) show polygons which *may* form part of a pattern.

Note that, in plane models representing patterns on compact surfaces, '*crosses*' indicate the vertices of the polygons forming the pattern, while '*dots*' denote the vertices of the 2n-gon which forms the compact surface.

Fig. 5.1.2 shows an example of a pattern on a torus *T*. We can see immediately that, on the space model of *T*, the region *A* is a 4-gon, that it meets its neighbours in

Vertices on the pattern are represented by crosses

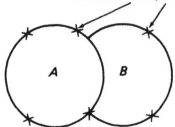

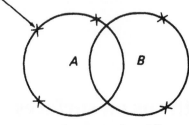

(a) Polygons meet along an
     incomplete edge

(b)   Polygons *A* and *B*
         overlap

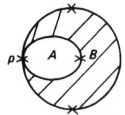

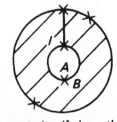

(c)   *B* meets itself at *p*

(d) *B* meets itself along the
     edge *l*

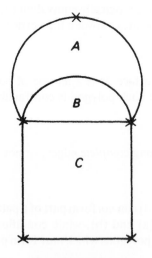

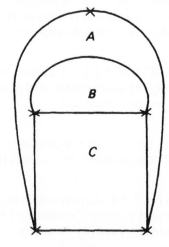

(e)   Possible pattern
        on a compact surface

(f)   Possible pattern
        on a compact surface

Fig. 5.1.1.

the correct way, and that it does not meet itself. However, a little more care is required when examining the remaining polygons. But it can be checked that they too satisfy conditions P1 and P2. For example, piercing together the region D, as we

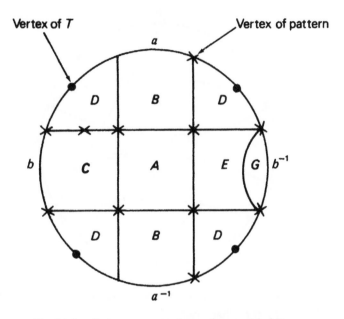

Fig. 5.1.2 — Pattern represented on a plane model of T.

have done in Fig. 5.1.3, we can see that D is an 8-gon meeting its neighbours in the correct way on the space model of T.

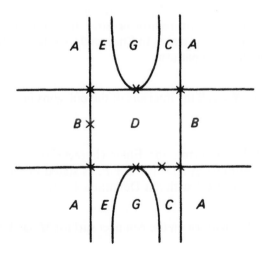

Fig. 5.1.3.

As we can see in Fig. 5.1.4, with the polygons A and B, it is usually quite easy to spot on the plane model when a polygon fails to satisfy condition P2. For readers who

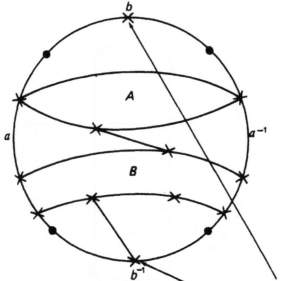

These *two* vertices represent *one* vertex on the space model; similarly for the other cases

Fig. 5.1.4.

find this difficult, we suggest that they cut out the plane model, construct a space model, and examine the mosaic of polygons on the space model, so checking conditions P1 and P2 directly. After a little practice, they should be able to spot what is happening at the plane model stage. In Chapter 9 we return to this, and provide a useful technique for checking patterns.

DEFINITION 5.1.3   We call the interiors of the polygons in a pattern the **faces** of the pattern.

In the standard development, the Euler characteristic is defined by way of patterns, giving the formula $\chi(M) = V - E + F$, as stated in section 4.5. Our next theorem shows that this agrees with our Definition 4.1.1.

THEOREM 5.5.1   If a pattern on the compact surface M has V vertices, E edges, and F faces, then

$$\chi(M) = V - E + F .$$

PROOF First we modify the pattern, if necessary, so that the vertices of M represent vertices of the pattern. (Remember that more than one vertex of M may give just *one* point on the space model of M; and similarly more than one vertex of the pattern represented on the plane model of M may give just *one* vertex of the mosaic of polygons on the space model of M.)

For example, in the pattern of Fig. 5.1.2, we could move an edge of the polygon B so that it passed through a vertex of T, and then we could introduce a new vertex at this point. This would leave us with the pattern shown in Fig. 5.1.5.

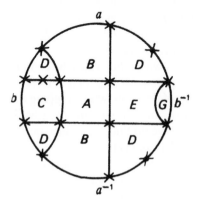

Fig. 5.1.5.

NOTE 5.1.1 *Notice* that, *on the plane model*, B appears in two distinct pieces and the vertex just introduced is represented by *four* vertices or 'crosses', as shown. Moreover, by involving all 2n vertices of the plane model of M in this way, the 2n vertices of the plane model of M form not more than n vertices of the new mosaic of polygons on the space model of M, i.e. the new pattern on M.

In all that follows, it should be remembered that when we talk about vertices, edges and faces of the pattern, we have in mind the *space model* of M, *unless otherwise stated*.

If we do introduce a new vertex in the way illustrated in Fig. 5.1.5, we notice that the number V-E + F is the *same* for the two patterns in Figs. 5.1.2 and 5.1.5. We increase the number of vertices by one, but in doing so, we divide an edge of the pattern into two parts, so producing an extra edge. The number of faces is unchanged, hence $V - E + F$ remains the same.

We now embark on a sequence of adjustments to the pattern on M. Each adjustment leaves the number $V - E + F$ unchanged.

First, wherever an edge of the pattern '*crosses*' an edge of M, when the plane model is made up into a space model, we introduce a new vertex. Every time we do this, we divide an edge of the pattern. As argued before, V and E increase by one, and $V - E + F$ remains the same.

Next, wherever a face of the new pattern '*crosses*' an edge of M, on the plane model, we introduce a new edge. Every time we do this, we divide a face in the pattern into two parts, so producing an extra face. Thus both E and F increase by one, while V remains unchanged. Hence $V - E + F$ remains the same.

We have now ensured that, on the plane model, the vertices of M are completely covered by vertices of the pattern, and that the edges of M are completely covered by edges of the pattern.

Now we begin to remove the pattern (as represented on a plane model) from the interior of the plane model of M in the following way. We choose a vertex p inside M, and we notice that, by P2, this vertex meets exactly as many faces as it meets edges, say m. Thus, when we delete p and each of the edges on which p lies, we replace m faces by a single face. Thus we reduce F by $m - 1$, E by m, and V by 1. On balance, the number $V - E + F$ remains the same.

As we can see in Fig. 5.1.6, which illustrates the complete process for a pattern on T, we may no longer have a pattern at this stage. However, this does not prevent us from continuing the process of deletion, while keeping the number $V - E + F$ fixed.

The next step is to remove all vertices lying on a single edge, and to remove the associated edges. Every time we do this, we reduce both V and E by one, while F is unchanged. Hence $V - E + F$ remains the same.

By repeating this whole process, we eventually remove all the vertices inside M on the plane model, while $V - E + F$ remains fixed.

However, as Fig. 5.1.6(e) shows, we may still be left with some edges inside M. But every time we delete such an edge, we combine two regions, hence losing a face. Thus, once again, $V - E + F$ remains the same.

Finally, we reach a stage where all vertices and edges in the space model are represented by vertices and edges lying on the boundary of M in the plane model, as shown in Fig. 5.1.6(f). From now on, every time we remove a vertex in the space model, we are actually removing the corresponding set of vertices on the boundary of M in the plane model, and two edges in the space model become one edge. However, the number of faces remains just one. Hence $V - E + F$ is unchanged. Eventually, if we remove all vertices which do not lie on the vertices of the 2n-gon M, we are left with the plane model of M. The 2n edges of M give n edges in the space model of M, and the 2n vertices of M give v vertices in the space model. Thus $V - E + F = v - n + 1$. But the left-hand side of this equation is the same as the original $V - E + F$, while the right-hand side is our definition of $\chi(M)$. Thus $V - E + F = \chi(M)$, as required.

The fact that the number $V - E + F$, associated with a pattern on a compact surface is entirely dependent on the surface turns out to be invaluable. Theorem 5.1.1 plays a prominent part in the remainder of this book and helps us to discover many interesting properties of surfaces.

5.2 COMPLEXES

A certain kind of pattern on a compact surface is of particular importance; it is called a 'complex'. Formally we have

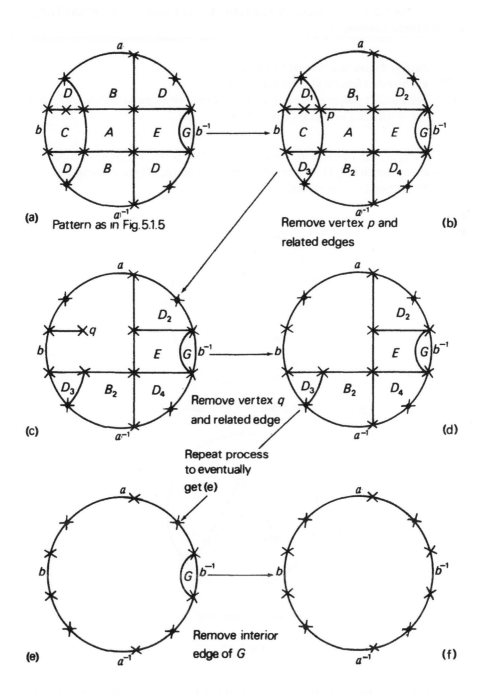

(a)   Pattern as in Fig.5.1.5

Remove vertex *p* and   (b)
related edges

(c)

Remove vertex *q*
and related edge   (d)

Repeat process
to eventually
get (e)

(e)

Remove interior
edge of *G*   (f)

Fig. 5.1.6.

DEFINITION 5.2.1   A **complex** is a pattern on a compact surface M satisfying the following three conditions:

C1   each polygon has at least three edges,
C2   each vertex lies on at least three edges,
C3   two distinct polygons meet at a single vertex, or along a single edge including the associated vertices, or not at all.

For example, the pattern shown in Fig. 5.2.1(a) is *not* a complex on S, since it violates every condition. However, it becomes a complex if we remove the edges l and m and the vertex x, as shown in Fig. 5.2.1(b).

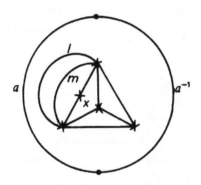

(a)   *Not* a complex on a sphere S

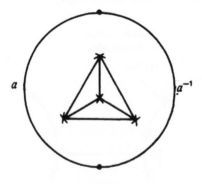

(b)   A complex on S

Fig. 5.2.1.

Let C be a complex on the compact surface M. Let F_n be the number of faces having exactly n edges (n⩾3). Let V_n be the number of vertices lying on exactly n edges (n⩾3). Then

$$F = F_3 + F_4 + F_5 + \ldots\ldots$$

$$V = V_3 + V_4 + V_5 + \ldots\ldots$$

Since each edge belongs to exactly two polygons in the space model of M, by counting the edges face by face, we obtain the equation

(1) $2E = 3F_3 + 4F_4 + 5F_5 + \ldots\ldots$

Similarly, by counting the edges vertex by vertex, we obtain the equation

(2) $2E = 3V_3 + 4V_4 + 5V_5 + \ldots\ldots$

We can now express the equation

$$\chi(M) = V - E + F$$

in the form

(3) $2\chi(M) = (2V_3 + 2V_4 + 2V_5 + \ldots\ldots) - (F_3 + 2F_4 + 3F_5 + \ldots\ldots)$

or

(4) $2\chi(M) = (2F_3 + 2F_4 + 2F_5 + \ldots\ldots) - (V_3 + 2V_4 + 3V_5 + \ldots\ldots)$

Adding equations (3) and (4), we obtain $4\chi(M) = (V_3 - V_5 - 2V_6 - 3V_7 - 4V_8 - \ldots\ldots) + (F_3 - F_5 - 2F_6 - 3F_7 - 4F_8 - \ldots\ldots)$ from which we deduce, since $F_n \geqslant 0$ and $V_n \geqslant 0$ for each n, the following results.

(a) If M is a sphere or a projective plane, so that $\chi(M) > 0$, then *any* complex on M must
    either contain some triangular faces
    or contain some trivalent vertices (i.e. vertices lying on exactly 3 edges).
(b) If M is a torus or a Klein bottle, so that $\chi(M) = 0$, then *any* complex on M
    either contains only quadrilateral faces and each vertex lies on precisely four edges.
    or contains some triangular faces or some trivalent vertices.
(c) If M is *not* a basic surface, so that $\chi(M) < 0$, then *any* complex on M must contain a polygon with at least five edges *or* a vertex lying on at least five edges.

Many properties of general complexes on compact surfaces may be deduced by manipulating equations (3) and (4) in a suitable way.

EXAMPLE 5.2.1   Show that every complex on a projective plane contains at least one of the following:

a triangle,   a quadrilateral,   a pentagon.

Solution   In this case, equations (3) and (4) become

$$2 = (2V_3 + 2V_4 + 2V_5 + \dots) - (F_3 + 2F_4 + 3F_5 + \dots)$$

and

$$2 = (2F_3 + 2F_4 + 2F_5 + \dots) - (V_3 + 2V_4 + 3V_5 + \dots) .$$

Eliminating F_6, we get

$$(3F_3 + 2F_4 + F_5 - F_7 - 2F_8 \dots) - (2V_4 + 4V_5 + 6V_6 + \dots) = 6 .$$

Hence, at least one of the numbers F_3, F_4, F_5 must be *non-zero*. This gives the required result.

Notice that in equations (1) and (2), and in equations (3) and (4), the roles of V_n and F_n ($n \geqslant 3$) are *interchanged*. This leads us to the concept of '*dual complexes*', which we define as follows.

DEFINITION 5.2.2   The complexes C and C' on the compact surface M are called **dual complexes** if $E = E'$, $V_n = F'_n$ ($n \geqslant 3$) and $F_n = V'_n$ ($n \geqslant 3$).

To find a dual complex C' for a given complex C on the space model of the compact surface M, we simply place a vertex for C' within each face of C and then join two such vertrices with an edge in C', whenever the corresponding faces of C meet along an edge in C. This process is illustrated in Fig. 5.2.2, where dual complexes are shown on the sphere S.

Naturally, the dual of the dual of the complex C on M is C itself.

5.3   REGULAR COMPLEXES

In this section, we consider a special kind of complex defined as follows.

DEFINITION 5.3.1   **A regular complex** on a compact surface M is a complex in which each polygon has the same number of edges and each vertex lies on the same number of edges. If a regular complex on M consists of a-gons and has b-valent vertices (i.e. each vertex lies on b edges), then we denote the complex by $(a,b)M$.

Note that in Fig. 5.2.2, we see examples of $(4,3)S$ and $(3,4)S$ on the plane model of S.

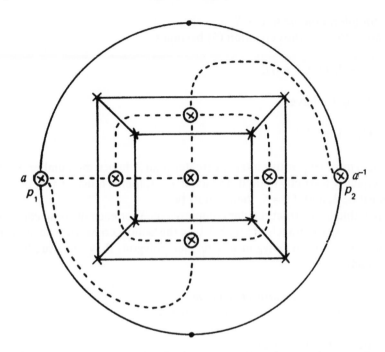

Fig. 5.2.2 — Dual complexes on a sphere S, where one complex is shown by the broken line with vertices $\otimes$. The vertices p_1 and p_2 on the boundary of the *plane* model of S form just *one* vertex on the *space* model of S.

For $(a,b)M$, we have $F = F_a$ and $V = V_b$; hence equations (1) and (2) of section 5.2 become

(1') $2E = aF$,
(2') $2E = bV$.

Substituting in the equation

$$\chi(M) = V - E + F ,$$

we obtain

$$\chi(M) = 2E/b - E + 2E/a .$$

Hence

(5) $\chi(M)/2E = 1/a + 1/b - 1/2$.

Using equation (5), we can determine all possible regular complexes on the basic surfaces, as follows.

Regular complexes on the sphere S
In this case, $\chi(S) = 2$, thus equation (5) becomes

(5') $1/a + 1/b = 1/2 + 1/E$.

Since $E > 0$, we have

(6) $1/a + 1/b > 1/2$.

Now a and b must be positive integers greater than 2, hence the number of solutions to this inequality is strictly limited. In fact, if a = 3, then b must be 3, 4, or 5; if a = 4, then b must be 3; and if a = 5, then b must be 3.

It is possible to contruct regular complexes on S for each of these combinations, as shown in the space models of Fig. 5.3.1. In the 'solid' geometric form in which they are shown, where each polygon is a regular polygon, they are usually called the **Platonic solids**.

Regular complexes on the projective plane
In this case, $\chi(P) = 1$ and equation (5) becomes

(5") $1/a + 1/b = 1/2 + 1/2E$.

Thus a and b must again satisfy the inequality (6), for which the possible solutions are:

$$a = 3 \quad \text{and} \quad b = 3, 4, \text{or } 5$$
$$b = 3 \quad \text{and} \quad a = 3, 4, \text{or } 5 \ .$$

However, if we assume $(3,3)P$ does exist, then from equation (5"), we find that $E = 3$. Then from equation (2'), we get $V = 2$. But we cannot form a complex of triangles with just two vertices. Hence we cannot form a regular complex on P with a = b = 3.

If we assume that $(4,3)P$ exists, then $E = 6$, $F = 3$, and $V = 4$. Clearly it is not possible to fit together three squares on any compact surface in such a way that each edge meets two squares and each pair of squares meet at most along one edge. Thus $(4,3)P$ is *not* possible. Of course, this means that the dual complex $(3,4)P$ does *not* exist.

The remaining possibilities $(3,5)P$ and its dual $(5,3)P$ can be realized, as shown as plane models in Fig. 5.3.2. In this case, the number of edges is 15 by equation (5").

Regular complexes on the torus and Klein bottle
For each of these surfaces $\chi(M) = 0$, therefore equation (5) becomes

(5''') $1/a + 1/b = 1/2$.

This time the possible solutions are

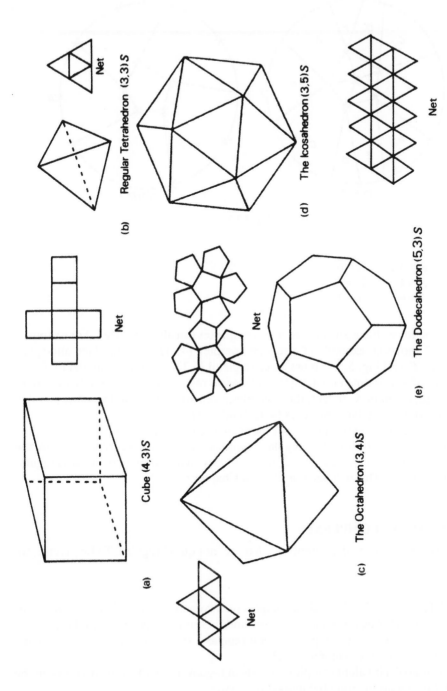

Fig. 5.3.1 — The cube and the octahedron are dual complexes on the sphere. The icosahedron and the dodecahedron are dual complexes. The regular tetrahedron is self-dual.

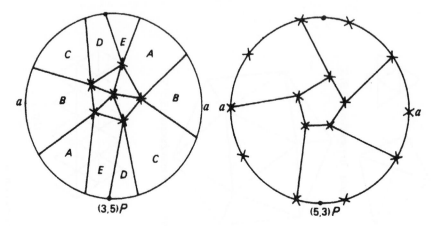

Fig. 5.3.2.

$$a = 3 , \quad b = 6$$
$$a = 4 , \quad b = 4$$
$$a = 6 , \quad b = 3 .$$

For the torus, the associated regular complexes can all be realized. For example, Fig. 5.3.3 shows an example of a regular complex $(4,4)T$ on both plane and space models of T, and Fig. 5.3.4 shows an example of $(6,3)T$ on a plane model of T.

Notice that this time equation$(5''')$ places no restriction on the number of edges in the regular complex. In fact, the examples given are just two of the many possible examples of the regular complexes $(4,4)T$ and $(6,3)T$.

In Fig. 5.3.5, we show another of the regular complexes $(4,4)T$.

For the Klein bottle, the regular complex $(4,4)K$ can be realized (see the exercises). However, the case $a = 3$, $b = 6$, and its dual cannot be realized as regular complexes on K. This will become clear in Chapter 6.

5.4   b-VALENT COMPLEXES

In this section we consider complexes that are 'not quite' regular. To be precise, we give:

DEFINITION 5.4.1   A **b-valent complex** on a compact surface M is a complex for which each vertex lies on exactly b edges and therefore meets exactly b faces.

We could call such complexes '**semi-regular**', since they satisfy a half of the regularity condition of Definition 5.3.1.

The **dual** of a b-valent complex consists of b-gons, but with *no* restrictions on the number of b-gons meeting the individual vertices.

For a b-valent complex, $2E = bV$, hence the equation $\chi(M) = V - E + F$ takes the form

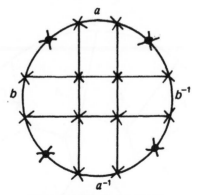

(a)   Plane model of an example
       of (4,4)T

(b)   Space model of the (4,4)T
       shown in (a)

Fig. 5.3.3.

$$\chi(M) = 2E/b - E + (F_3 + F_4 + F_5 + \ldots) .$$

Using equation (1), we get

(7) $\chi(M) = (1/b - 1/2)(3F_3 + 4F_4 + 5F_5 + \ldots) + (F_3 + F_4 + F_5 + \ldots) .$

Taking the special case b = 3, we obtain the equation

(8) $6\chi(M) = 3F_3 + 2F_4 + F_5 - F_7 - 2F_8 - \ldots$

from which we deduce the following facts.

(a) *Any* trivalent complex on S or P must contain at least one of the following:
    a triangle, a quadrilateral, a pentagon.
(b) *Any* trivalent complex on K or T
        either contains *only* hexagons,
        or contains at least one of the following:
        a triangle, a quadrilateral, a pentagon.

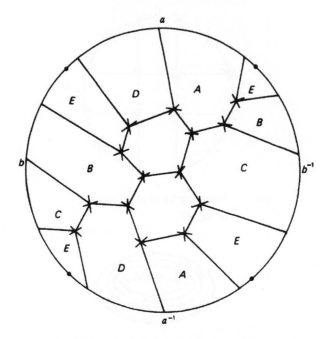

Fig. 5.3.4 — Plane model of an example of $(6,3)T$.

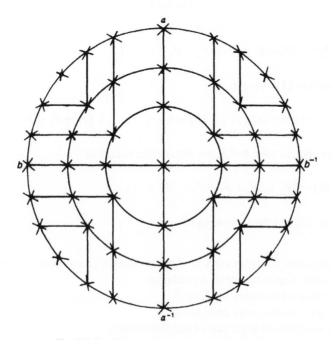

Fig. 5.3.5 — Plane model of an example of $(4,4)T$.

DEFINITION 5.4.2   The dual of a trivalent complex on the compact surface M is called a **triangulation** of M.

In the standard treatment of surface topology, triangulations play a central role.

Many more specific properties of trivalent complexes, and therefore also of triangulations, may be deduced from equation (8). You may like to do this for yourself. For example, for *any* trivalent complex on S consisting of triangles and octagons, we have $3F_3 - 2F_8 = 12$.

As an exercise, check that this property holds for the trivalent complex on S obtained by shaving the corners from a cube, as shown, as a space model, in Fig. 5.4.1.

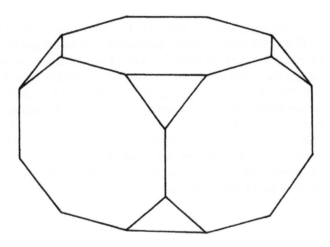

Fig. 5.4.1 — An example of a trivalent complex on S.

5.5   COMMENTS

In this chapter, we have introduced techniques which can be used to discover properties of patterns on compact surfaces. Some of these properties we have discussed, but many others await discovery. We invite the reader to uncover some of these for himself, or herself, by following the ideas put forward here.

We view patterns on surfaces as topologists. For us, there is no difference between the pattern formed by the edges of a cube and the pattern on a sphere shown, as space models, in Fig. 5.5.1.

However, for a geometer, there is a great deal of difference. Much is known, from a geometric standpoint, about patterns drawn on a sphere. From this point of view, they are regarded as *rigid* patterns. As such, of course, they have a much richer structure than when considered topologically. Excellent books by Wenninger [31] and Coxeter [5] pursue this line of thought.

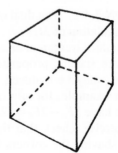

Fig. 5.5.1.

The study of solid figures goes back at least to the time of the ancient Greeks some 2500 years ago. Plato (4th century BC) complained that investigators of the solid figures did not receive due support from the state. The problem is not new!

Plato and his associates knew that there could be at most *five* regular polyhedra. i.e. all faces are the same regular polygon. But the proof is probably due to Theaetetus. For **uniform** polyhedra, which have the same arrangement of regular polygons at every vertex with just two faces meeting at any edge, the list given by Coxeter and others was shown to be complete, only recently, by John Skilling [29].

It has been said that the most famous textbook of all, Euclid's *Elements*, was not in fact written as a textbook, but rather as an introduction to the five regular solids known to the ancient world as the Platonic solids.

5.6 EXERCISES

(1) Determine which of the diagrams in Fig. 5.6.1 represent

  (a) a pattern,
  (b) a complex,

on the given surface.

In each case, evaluate $V - E + F$ and compare with the Euler characteristic of the surface.

(2) Sketch a space model of S which illustrates the complex shown in Fig. 5.6.2 in a more familar form.

(3) Trace Fig. 5.6.3. Cut out the diagrams, fold along the lines, and plait them to form two examples of triangulations of S. Count the vertices, edges and faces, and check the formula $V - E + F = \chi(S)$.

(4) (a) Trace the nets shown in Fig. 5.3.1 and use them to construct examples of the platonic solids.

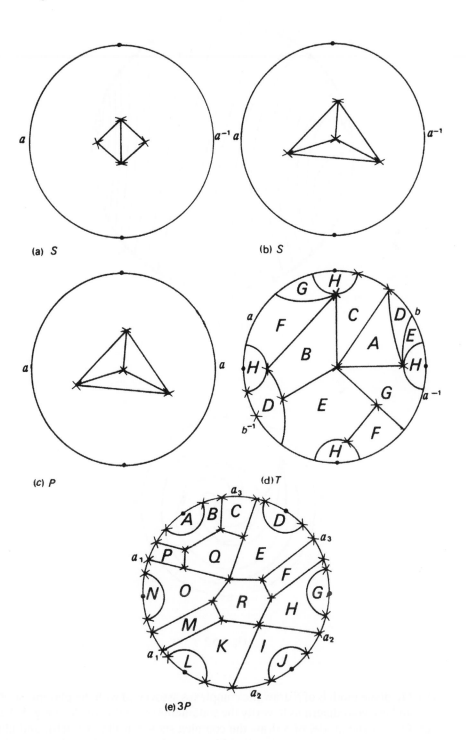

(a) *S*

(b) *S*

(c) *P*

(d) *T*

(e) 3*P*

Fig. 5.6.1.

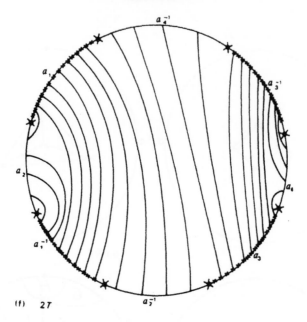

Fig. 5.6.1.

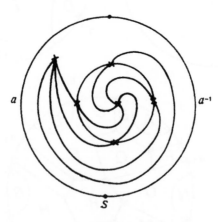

Fig. 5.6.2.

(b) On plane models of S draw the complexes associated with the platonic solids and use your diagrams to verify the statements on duality made in Fig. 5.3.1.

(c) On a plane model of S draw the complex shown in Fig. 5.6.4(b), and glue these together to form the rhombitruncated cuboctahedron shown in Fig.

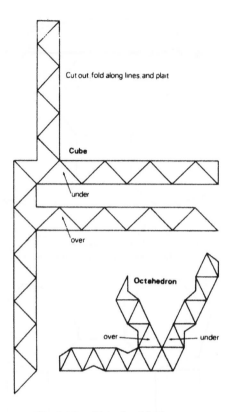

Fig. 5.6.3 — Plaited polyhedra.

5.6.4(a). Draw this complex on a plane model of S, and by counting V, E, and F, check the formula $V - E + F = \chi(S)$.

(5) Draw the plane models corresponding to the 'toroidal polyhedra' shown in Fig. 5.6.5 and in each case check the formula $V - E + F = \chi(M)$.

(6) (a) Find the solutions of the equation

$$1/a + 1/b = 1/2 + 1/E$$

when a or b is allowed to equal 2, and thus find the possible degenerate regular spherical polyhedra. Sketch them.

(b) Show that only the sphere and the projective plane can have degenerate regular polyhedra.

(7) On a plane model of K, sketch an example of $(4,4)K$.

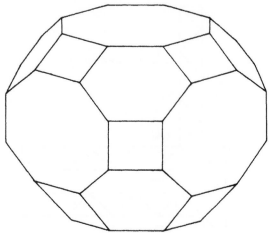

(a)   Rhombitruncated cuboctahedron

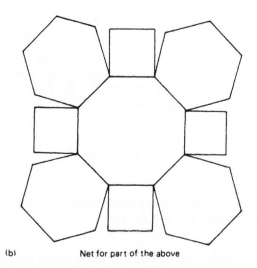

(b)   Net for part of the above

Fig. 5.6.4.

(8) (a) On a plane model of T, sketch an example of $(6,3)T$ having more faces than
        the example shown in Fig. 5.3.4.
    (b) Draw two different examples of $(3,6)T$ on plane models of T.
    (c) Sketch a space model of an example of $(3,6)T$.

(9) (a) Prove that for any complex on a surface M the number of polygons with an
        odd number of edges is even. What is the dual statement? (That is, what does
        this say about the dual of any complex on M.)
    (b) Write down the duals of the statements (a) and (b) section 5.4.

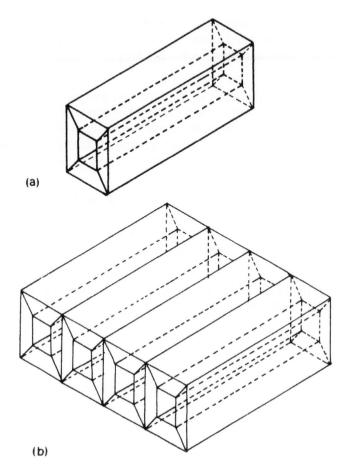

(a)

(b)

Fig. 5.6.5 — Toroidal polyhedra.

(10) Prove that *no* triangulation of T exists for which each vertex meets exactly k triangles, (k = 3, 4, 5).

(11) The plane model of $500P$ is triangulated in such a way that each edge of the polygon is divided into three edges in the triangulation, and the interior of the polygon contains 1000 vertices of the triangulation. Determine the excess of the number of edges interior to the polygon over the number of triangles.

(12) (a) Show that if a complex on S has no triangles or quadrilateral faces, then it has at least 12 pentagonal faces. Is there a complex on S with the minimum number of pentagons?

    (b) If a trivalent complex on S is made up from pentagons, septagons and exactly six quadrilaterals, show that there are exactly as many pentagons as septagons. Sketch a space model describing such as complex.

(c) If a trivalent complex on T is made up from quadrilaterals and octagons, show that there are exactly as many quadrilaterals as octagons. Sketch a space model describing such a complex.

(13) A complex C on P is made up from triangles and pentagons, and has trivalent, four-valent, and five-valent vertices only. If C has exactly five four-valent vertices and one five-valent vertex, and if the number of triangles differs from the number of pentagons by at most three, find the number of triangles and the number of pentagons. Sketch an example of the complex C on a plane model of P.

(14) Show that for any b-valent complex on K or T made up from triangles, quadrilaterals, and pentagons *either* there are exactly as many triangles as pentagons, *or* there are at least twice as many triangles as quadrilaterals and there are at least five times as many triangles as pentagons.

(15) On the projective plane determine all possible trivalent complexes made up from an equal number of m-gons and n-gons.

6

Maps and graphs

6.1 COLOURING MAPS ON SURFACES

The proof, or disproof, of the four-colour conjecture engaged the attention of many mathematicians over a long period. In fact, the problem succumbed to modern techniques only recently, when a proof was found with the considerable help of a computer. Basically, the conjecture states that *any* pattern on a sphere can be coloured with at most four colours, and, for some patterns, *all four* colours are required.

In this section, we consider the problem of colouring patterns on *any* compact surface. But first we need some definitions, including one that tells us what it means to 'colour' a pattern.

DEFINITION 6.1.1   A **map** on a compact surface M is a pattern on M, where each vertex lies on *at least three* edges.

DEFINITION 6.1.2   To **colour** a map on the compact surface M is to associate a colour with each face of the map in such a way that if two distinct faces have a common edge, then the two faces are associated with different colours. A map is **N-colourable**, if it can be coloured with N colours.

The basic question we ask is:

'*What is the smallest number of colours needed to colour any map on the compact surface M?*'

For a regular complex on the compact surface M, with E edges and F n-gon faces, we have the equation $2E = nF$. Thus $2E/F = n$ is the number of edges per face. In a general map, the number $a = 2E/F$ represents the *average* number of edges per face. Not surprisingly, this number 'a' plays an important part in the solution of map colouring problems.

LEMMA 6.1.1   If $a < N$ for *all* maps on the compact surface M, where $a = 2E/F$ and N is some fixed positive integer, then N colours are sufficient to colour all maps on M.

PROOF   The proof is by induction on F, the number of faces of the map H on the compact surface M.

If $F \leq N$, then a different colour can be associated with each face of M. This starts the inductive process.

Now let us suppose that every map on M with k faces can be coloured with N colours.

Consider the map H on M with k + 1 faces. Since a < N for this map, at least one face, say A, has fewer than N edges. Let us carry out the following construction.

Place a new vertex within the face A. Link this vertex to each of the vertices of the polygon formed by the edges of A. Delete all edges of this polygon, and then delete all vertices which lie on only two edges. The result is a new map H' on the compact surface M, with only k faces. This process is illustrated in Fig. 6.1.1.

By the induction hypothesis, the map H' can be coloured with N colours.

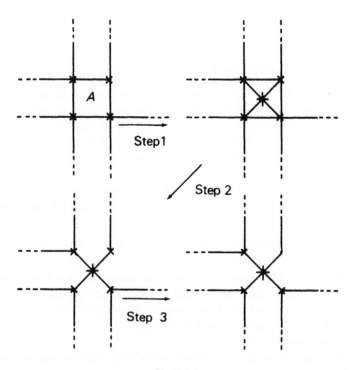

Fig. 6.1.1.

Next, we reverse the above procedure, and reintroduce the face A. Since A meets fewer than N faces of H (recall that A was selected as a face with less than N edges), there is at least one colour which can be used to colour A so as to give us a colouring of H. Thus H can be coloured with N colours. An appeal to induction completes the proof.

Lemma 6.1.1 relates the number N, of colours required to colour maps on the compact surface M, to the numbers a $= 2E/F$, whose values are determined by the maps themselves.

But what we are really looking for is a direct link between N and the compact surface M, itself. We can get this by relating 'a' to $\chi(M)$.

Since each vertex in a map lies on at least 3 edges, we know that $3V \leqslant 2E$. Thus

$$\chi(M) = V - E + F$$
$$\leqslant 2E/3 - E + F.$$

Hence

$$E \leqslant 3(F - \chi(M)).$$

Thus

(1) $a = 2E/F \leqslant 6(1 - \chi(M)/F).$

If $\chi(M) > 0$, then $a \leqslant 6(1 - \chi(M)/F) < 6$.

Thus Lemma 6.1.1 immediately implies:

THEOREM 6.1.1   Six colours are sufficient to colour *any* map on the sphere S or the projective plane P.

Of course, for the sphere, this is a very weak result. However, for the projective plane it is *exact*. The map shown in Fig. 6.1.2, which needs exactly six colours, proves this.

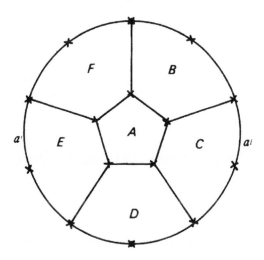

Fig. 6.1.2.

If $\chi(M) = 0$, then the inequality (1) becomes $a \leqslant 6 < 7$. In this case, Lemma 6.1.1 implies:

THEOREM 6.1.2   Seven colours are sufficient to colour *any* map on the torus or the Klein bottle.

These results are anolgous to those for the sphere and projective plane, in that, for the Klein bottle K, this is a weak result, but, for the torus T, it is exact. This time, the *orientable* compact surface T has the exact result, whereas in Theorem 6.1.1, it is the *non*-orientable projective plane which has the exact result.

Fig. 6.1.3 shows that there is a map on the torus requiring *all* 7 colours.

Notice that we have seen this complex on T before. It is, in fact, $(6,3)T$, as shown in Fig. 5.3.4.

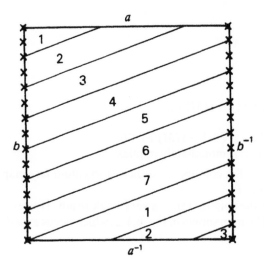

Fig. 6.1.3.

If $\chi(M) < 0$, and H is a map on M, then we can assume that H has at least $N + 1$ faces, otherwise H is obviously N-colourable. In this case, we have

$$1/F \leqslant 1/(N + 1) \ .$$

Hence $\chi(M)/(N + 1) \leqslant \chi(M)/F$, since $\chi(M)$ is negative. Using this in the inequality (1), we obtain

$$a \leqslant 6(1 - \chi(M)/(N + 1)) \ .$$

Thus, if $6(1 - \chi(M)/(N + 1)) < N$, then Lemma 6.1.1 implies that each map on M is N-colourable.

This latter inequality can be written:

$$N^2 - 5N + 6\chi(M) - 6 > 0 \ .$$

The quadratic $x^2 - 5x + 6\chi(M) - 6$, whose graph is shown in Fig. 6.1.4, has roots $(5 \pm \sqrt{49 - 24\chi(M)})/2$.

Notation: we denote the largest *integer* less than or equal to the real number k by [k].
   Thus, if we take N to be the smallest integer greater than

$$(5 + \sqrt{49 - 24\chi(M)})/2 ,$$

then

$$N = [1 + (5 + \sqrt{49 - 24\chi(M)})/2],$$

and

$$6(1 - \chi(M)/(N + 1)) < N.$$

   We have proved

THEOREM 6.1.3   For the compact surface M, let $\chi(M) < 0$, then *any* map on M is N_M-colourable, where

$$N_M = [7 + \sqrt{49 - 24\chi(M)})/2] .$$

   This theorem was proved by Heawood in 1890, and was followed by

Heawood's Conjecture:
For *any* compact surface M, N_M is the *minimum* number of colours needed to colour *all* maps on M.
   Notice that, although we have obtained this formula for compact surfaces M for which $\chi(M) < 0$, it gives reasonable values for the other compact surfaces, as we can see in Table 6.1.1.
   It has been found that only *six* colours are needed to colour all maps on a Klein bottle. But in *all* other cases, Heawood's conjecture has now been verified.

6.2   EMBEDDING GRAPHS IN SURFACES

DEFINITION 6.2.1   A **graph** G in a surface M is a finite set of points in M, called the **vertices** of G, and a finite set of lines in M, called the **edges** of G, which link certain pairs of distinct vertices in such a way that any pair of distinct vertices determines *at most one* edge.

DEFINITION 6.2.2   We say that the graph G is **embedded** in M, when it is drawn so that *no* two edges meet, *except* at a vertex.

DEFINITION 6.2.3   The graphs G_1 and G_2 are said to be **isomorphic** if their vertices can be placed in one–one correspondence in such a way that a pair of vertices in G_1 are linked by an edge if and only if the corresponding pair of vertices in G_2 are linked by an edge.

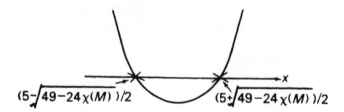

$$(5-\sqrt{49-24\chi(M)})/2 \qquad\qquad (5+\sqrt{49-24\chi(M)})/2$$

Fig. 6.1.4.

Table 6.1.1

$\chi(M)$	2	1	0	-1	-2	-3	-4	-5	-6	-7	-8	....
N_M	4	6	7	7	8	9	9	10	10	10	11	....

For example, the graphs shown in Fig. 6.2.1 are isomorphic. In Fig. 6.2.1(b), the graph is embedded in a torus T, while the graph in Fig. 6.2.1(a) is *not*. Remember that the graphs must be pictured on the *space model* of T.

Two kinds of graphs have special interest for us.

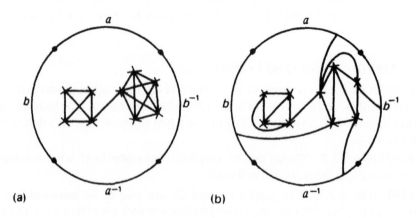

Fig. 6.2.1.

First, the **complete graph** on n vertices denoted by K_n. This graph has each vertex linked to every other vertex by an edge.

Second, the **complete bipartite graph**, on m and n vertices, denoted by $K_{m,n}$. this graph has disjoint sets of m and n vertices. Each vertex in one set is linked by an edge to every vertex in the other set.

Fig. 6.2.2(a) shows K_4 embedded in the sphere S and Fig. 6.2.2(b) shows $K_{2,3}$ embedded in S.

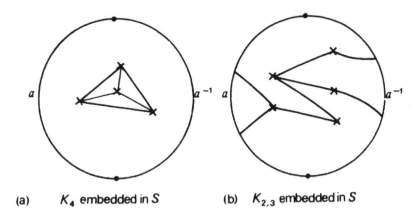

(a) K_4 embedded in S (b) $K_{2,3}$ embedded in S

Fig. 6.2.2.

Any graph can be embedded in *some* orientable compact surface. This can be seen by constructing a wire model of the graph, connecting any separate parts of the graph so that a compact surface is formed, and then drawing the graph on the surface of the wire so that the edges of the graph meet only at vertices.

Fig. 6.2.3 shows K_4 embedded in this way in its wire model.

Fig. 6.2.3.

Of course, this embedding is very wasteful in the sense that we have far more 'holes' in the surface than we actually need. In fact, as we saw in Fig. 6.2.2(a), we can embed K_4 in a compact surface with no 'holes' at all, i.e. a sphere.

This train of thought leads us to make the following definitions.

DEFINITION 6.2.4   Let the graph G be embedded in the *orientable* compact surface M. Let $\gamma(G)$ be the *maximum* value of $\chi(M)$ for *all* such orientable compact surfaces M. The even integer $\gamma(G)$ is called the **characteristic of** G.

DEFINITION 6.2.5   Let G be embedded in the orientable compact surface M. We call this embedding a **minimal embedding** of G (i.e. minimal with respect to the number of 'holes' required, see Corollary 4.2.1(a)) if $\gamma(G) = \chi(M)$. Also M is then said to be a **minimal** surface for G.

For certain graphs, including those considered here, König proved an important result, which, for our purpose, we state as

THEOREM 6.2.1   If the connected graph G is minimally embedded in the orientable compact surface M, and if the graph has V vertices, E edges, and the embedding produces F faces, then

$$V - E + F = \chi(M) \ .$$

As with several other theorems quoted in this chapter, a proof of this theorem lies outside the scope of this book.

6.3   PLANAR GRAPHS

Notice that Fig. 6.2.2(a) shows that $\gamma(K_4) = 2$. Graphs, like K_4, whose characteristic is 2, are called planar graphs, since a graph can be embedded in a plane if and only if it can be embedded in a sphere S.

For future reference, let us state this formally as:

DEFINITION 6.3.1   We call G a **planar** graph if $\gamma(G) = 2$.

Since $\gamma(K_4) = 2$, we can see that $\gamma(K_n) = 2$ for $n \leqslant 4$. However, K_5 is *non*-planar, as we can see, with the help of Theorem 6.2.1, in the following way.

If we suppose that K_5 can be embedded in a sphere, then such an embedding must be minimal, and, by theorem 6.2.1, if the embedding produces F faces, then we know that K_5 has 5 vertices and 10 edges, and so, from $2 = \gamma(K_5) = \chi(S) = V - E + F$, we see that $F = 7$. Now in K_5 any two vertices are linked by a unique edge, therefore, each polygon in the pattern has at least 3 edges. Hence

$$3F \leqslant 2E \ .$$

However, $E = 10$ and $F = 7$, gives a contradiction. We conclude that K_5 must be non-planar.

In a similar way, we can prove that $K_{3,3}$ is non-planar (see the exercises).

One of the main theorems of graph theory tells us that, in a certain sense, we have now found all possible non-planar graphs. To explain what we mean by this, let us call G and H **homeomorphic graphs** if they are isomorphic, *or*, if each graph can be obtained from a graph J by the introduction of new vertices on the edges of J. This process is illustrated in Fig. 6.3.1, where we see that the graphs G and H can be obtained in this way from K_5. Thus G and H are homeomorphic.

We call the graph A a **subgraph** of the graph B, if each vertex of A is a vertex of B and each edge of A is an edge of B.

For example, in Fig. 6.3.2, we have two subgraphs of $K_{5,4}$, each containing a subgraph homeomorphic to $K_{3,3}$.

We can now state

THEOREM 6.3.1 (Kuratowski's Theorem)   A graph is planar if and only if it has *no* subgraph homeomorphic to K_5 or $K_{3,3}$.

6.4   OUTERPLANAR GRAPHS

A particular type of planar graph is of some interest. We give

DEFINITION 6.4.1   A graph is said to be **outerplanar**, if it can be embedded in a sphere S so that one of the faces in the embedding contains each of the vertices of the graph on its boundary.

Obviously, any outerplanar graph is planar. In order to obtain a Kuratowski type theorem, in this section, we look for those planar graphs which are not outerplanar.

Any embedding of K_4 in a sphere S essentially looks like the embedding in Fig. 6.2.2(a). Thus K_4 is a forbidden graph as far as outerplanarity is concerned. Similarly, any embedding of $K_{2,3}$ in S essentially takes the form shown in Fig. 6.4.1(a). Thus $K_{2,3}$ is also a forbidden graph for outerplanarity.

At this stage, we appear to meet a difficulty, for the graph $K_4 - x$ (i.e. K_4 with one edge x deleted) is outerplanar, as we can see from Fig. 6.4.1(b). But $K_4 - x$ is homeomorphic to the forbidden graph $K_{2,3}$.

Fortunately, this is a minor 'hiccup'. We can get round it and produce a theorem for outerplanarity similar to Kuratowski's theorem for planarity. We state it as

THEOREM 6.4.1   A graph is outerplanar if and only if it has *no* subgraph homeomorphic to K_4 or $K_{2,3}$, with the possible exception of $K_4 - x$.

6.5   EMBEDDING THE COMPLETE GRAPHS

As we have seen in section 6.3, K_5 and $K_{3,3}$ are non-planar. Hence $\gamma(K_5) \leqslant 0$ and $\gamma(K_{3,3}) \leqslant 0$. In Fig. 6.5.1, we show an embedding of $K_{3,3}$ in T. Thus $\gamma(K_{3,3}) = 0$.

Similarly, by drawing an embedding of K_5 in T (see the exercises), we prove that $\gamma(K_5) = 0$.

Thus K_5 and $K_{3,3}$ are *not* forbidden graphs for T.

What are the forbidden graphs for T?

It has been shown that for each orientable compact surface M, there exists a *finite* set of forbidden graphs, i.e. graphs that cannot be embedded in M. From this, we see it is *possible* to construct a theorem in the style of Kuratowski's theorem. However, up to the present moment, no-one has been able to actually *exhibit* a specific set of forbidden graphs for *any* orientable compact surface other than S!

Nevertheless, if we take an alternative view, and try to relate certain classes of graphs to their minimal orientable compact surfaces, some progress can be made. Let us be more specific.

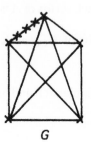 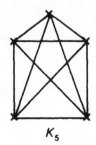 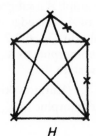

$$G \qquad\qquad K_5 \qquad\qquad H$$

Fig. 6.3.1.

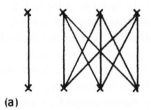 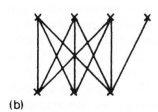

(a)   (b)

Fig. 6.3.2.

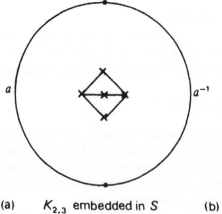 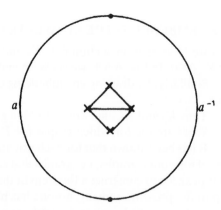

(a) $K_{2,3}$ embedded in S (b) $K_4 - x$ embedded in S

Fig. 6.4.1.

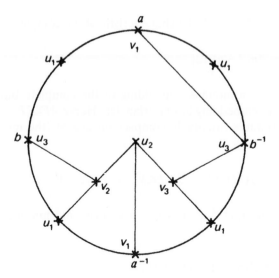

Fig. 6.5.1.

Let G be a graph with V vertices, E edges, and F faces, minimally embedded in the orientable compact surface M. Then

$$\gamma(G) = \chi(M) = V - E + F .$$

Since each polygon in the pattern formed by the graph has at least 3 edges, we know that $3F \leqslant 2E$. Hence

$$\gamma(G) \leqslant V - E + 2E/3 = V - E/3 .$$

Now $\gamma(G)$ is an even integer (see Corollary 4.2.1(a)), therefore, for *any* graph, we have:

$$\gamma(G) \leqslant 2[(V - E/3)/2],$$

using the notation introduced just prior to Theorem 6.1.3.

In particular, for the complete graph K_n, which has n vertices and $\binom{n}{2}$ edges, we obtain the inequality:

$$\gamma(K_n) \leqslant 2[(n - n(n - 1)/6)/2] = 2[n(7 - n)/12] .$$

To prove that we have equality here, we must, for each $n \geqslant 3$, display an embedding of K_n in an orientable compact surface M with Euler characteristic $\chi(M) = 2[n(7 - n)/12]$.

Certain values of n put up quite a struggle, but the last of them finally succumbed recently; see section 6.8. Thus, the following theorem was eventually proved.

THEOREM 6.5.1   For n$\geq$3, the characteristic of the complete graph K_n is:

$$\gamma(K_n) = 2[n(7 - n)/12] \; .$$

Consideration of a minimal embedding of the complete bipartite graph $K_{m,n}$ shows that such an embedding has *no* triangles. Hence $4F \leq 2E$.

Thus, for a suitable orientable compact surface M (the minimal surface), we have:

$$\gamma(K_{m,n}) = \chi(M) = V - E + F \leq V - E + E/2 = V - E/2 \; .$$

Bearing in mind that $K_{m,n}$ has $m + n$ vertices and mn edges, we deduce the inequality

$$\gamma(K_{m,n}) \leq 2[(m + n - mn/2)/2] \; .$$

The final embeddings of $K_{m,n}$ needed to prove equality in this case were found in 1965. We present the result as

THEOREM 6.5.2   For m, n$\geq$3, the characteristic of the complete bipartite graph $K_{m,n}$ is

$$\gamma(K_{m,n}) = 2[(m + n - mn/2)/2] \; .$$

6.6   SPROUTS

In this section and the next, we turn to somewhat lighter topics.

The problem of embedding graphs in surfaces can be converted into an interesting game in the following way.

The game of **sprouts** begins with n vertices placed on the compact surface M. A move consists of drawing a line in M, which joins one vertex to another vertex, or to itself, and then placing a new vertex on this line. There are just two rules.

(1) The line drawn must not cross a previously drawn line, cross itself, or pass *through* a vertex.

(2) No vertex may have more than three lines meeting it.

Two players take turns to move. The winner is the last player able to make a move.

Fig. 6.6.1 shows an example of a game played on a sphere S beginning with 3 vertices. Playing on S is equivalent to playing on a sheet of paper.

Initially, by rule (2), each vertex can have 3 lines drawn to meet it. Each move brings about 2 such meetings and introduces a new vertex, which may have just one more line drawn to meet it. Thus each move reduces, by one, the possible number of meetings that can be made between lines and vertices. It follows that the game of sprouts, on any compact surface, ends after at most $3n - 1$ moves.

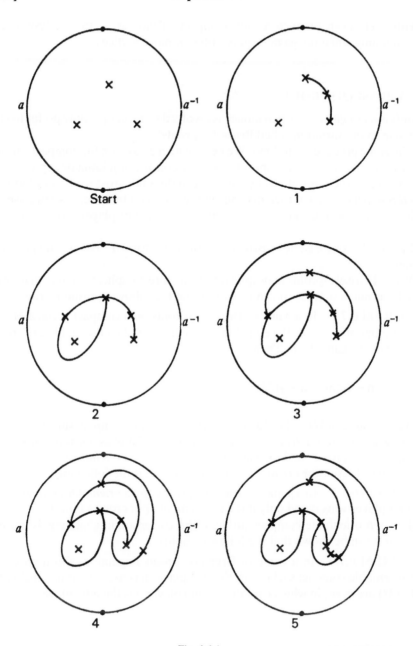

Fig. 6.6.1.

By examining all possible sequences of moves, it can be shown that, with two initial vertices on S, the second player has the advantage and by careful play should always win. Similarly, it can be shown that the first player has the advantage with 3, 4, or 5 initial vertices on S.

However, even on such a mild compact surface as S, the analysis becomes overwhelming when the game begins with 6 or more vertices.

6.7 BRUSSELS SPROUTS

Sprouts is a fair game, so when money is involved some people may prefer to change the game to the variation called **Brussels Sprouts**!

The game proceeds as follows. We begin with n *crosses* on the compact surface M. The arms of the cross are the sprouts. A move consists of joining two sprouts (which may belong to the *same* cross) by a line which does not cross itself, another line, or pass *through* a cross, and then placing another cross on this line. Two opposite arms of this latter cross must lie *along* the line. Again the last player able to move is the winner.

Fig. 6.7.1 shows a game of Brussels sprouts on a torus T. The second player is the winner.

While Brussels sprouts may appear to be a more complicated game than sprouts, in fact it is much simpler. It can be shown that the following result is true.

THEOREM 6.7.1   In a game of Brussels sprouts on a compact surface M, let the total number of moves to the end of the game be m, and let the number of crosses at the start of the game be n. Then

$$5n - 2 \leqslant m \leqslant 5n - \chi(M) \ ,$$

where, as usual, $\chi(M)$ is the Euler characteristic of the compact surface M.

For a sphere, when $\chi(M) = 2$, the total number of moves is *always* $5n - 2$. In our example above, n = 2, and the game ended after $5n - 2 = 8$ moves.

A general result for *orientable* compact surfaces is the following.

THEOREM 6.7.2   In a game of Brussels sprouts on an *orientable* compact surface, the first player wins if and only if the initial number of crosses is *odd*.

For non-orientable compact surfaces, the corresponding analysis is more difficult, but it can be shown that the following holds.

THEOREM 6.7.3   In a game of Brussels sprouts beginning with n crosses on a *non*-orientable compact surface M, the first player has the advantage, *unless both* n and $\chi(M)$ are *even*, in which case the second player has the advantage.

6.8 COMMENTS

The four-colour conjecture was made in 1852 by Francis Guthrie, although it is rumoured that Möbius was familiar with the problem in 1840. Over the years many mathematicians attempted a proof, and some published 'proofs' were accepted for a time, but were eventually seen to be wrong. Fortunately the work was not wasted, because such great efforts were directed towards solving the problem that several new developments in combinatorial topology and graph theory resulted.

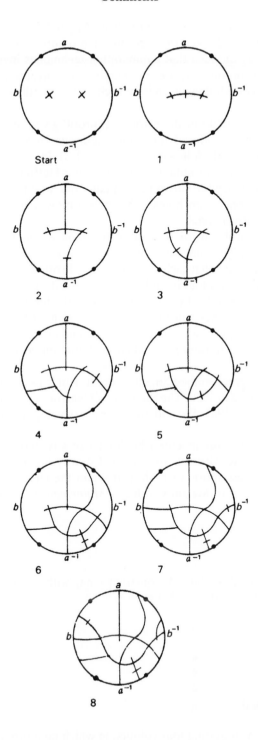

Fig. 6.7.1 — A game of Brussels sprouts on T.

Finally, the conjecture was settled in 1976 by Appel & Haken with the considerable help of a computer. In fact, this introduced into pure mathematics for the first time in a serious way the idea that enumeration techniques involving a computer could be used to give 'proofs'. In pure mathematics the conventional proof consists of a sequence of statements following on from each other according to the rules of some logic. (See [11].)

Heawood, who pointed out the error in a 'proof' given by Kempe of the four colour conjecture, later suffered the same experience himself over 'Heawood's conjecture' (see the remarks following Theorem 6.1.3). In 1890, he believed that he had proved equality in his formula, but one year later Heffter pointed out errors in the argument and managed to prove that equality did hold for orientable compact surfaces having from one to six 'holes'. The conjecture was finally settled by Ringel and Youngs as recently as 1968.

The embedding problem for the complete graph K_n is closely related to the Heawood conjecture. The connection is made by way of the observation that if a complex C has n faces such that each face meets $(n-1)$ other faces, then the complex needs exactly n colours to colour it, and the dual of C is the graph K_n. Heffter proved Theorem 6.5.1 for n = 8 to 12. In 1952, Ringel proved it for n = 13. Continued work up to 1968 by Ringel, Youngs, Gustin, Terry, and Welch settled all cases except for n = 18, 20, and 23. The proof was completed at the end of the sixties, by Jean Mayer, a professor of French Literature (!), when he found embeddings for these three values.

The theorems on graph embeddings are recent. Kuratowski's characterization of planar graphs appeared in 1930. The theorem of Chartrand and Harary on outerplanar graphs appeared in 1967. We still await similar developments on other compact surfaces.

The game 'sprouts' was invented by John Conway and Michael Paterson at Cambridge in 1967, and 'Brussels sprouts' by Conway slightly later, some say as a joke. An excellent description of both games and the circumstances surrounding their invention is given in Martin Gardner's *Mathematical Carnival* [12]; also see Giblin [13].

6.9 EXERCISES

(1) On a plane model describing A, construct a map with n faces such that each face meets every other face and such that each face has n edges.

A	n
Rectangle	4
Cylinder	4
Sphere	4
Möbius band	6

(2) Find a map on S, requiring four colours, in which no more than three faces are mutually adjoining.

(3) T needs seven colours to colour all maps.

   $2T$ needs eight colours to colour all maps.

   nT needs n + 6 colours to colour all maps.

True or false?

(4) Show that for any integer n, we can construct a 'three-dimensional map' (i.e. each country is a solid block) needing n colours to colour it.

(5) In Chapter 5 it was stated that the complexes $(3,6)K$ and $(6,3)K$ could not be realized, and that this would become clear in Chapter 6. What facts pointed out in Chapter 6 imply the non-existence of such complexes?

(6) If a≤b and a is an integer divisible by n, write down a sharper inequality relating a, b and n.

(7) (a) Show that the graph of Fig. 6.9.1 is planar. Do you recognize this graph?

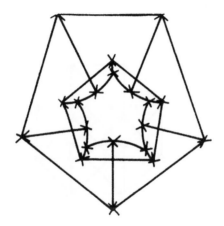

Fig. 6.9.1.

   (b) Show that the graphs of Fig. 6.9.2 are non-planar by using Theorem 6.3.1. Find minimal embeddings for these graphs in orientable compact surfaces.

(8) (a) Express the surface of a wire model of K_n in the form NT for some integer N.

   (b) Express the surface of a wire model of $K_{m,n}$ in the form NT for some integer N.

(9) Prove that $\gamma(K_{m,2}) = 2$ for each positive integer m.

(10) (a) Prove that $K_{3,3}$ is non-planar.

   (b) Deduce from Heawood's conjecture that it is not possible to embed K_{4+7n} in $4n^2T$ (n≥1).

(11) (a) Using Theorem 6.5.1 find $\gamma(K_n)$ for values of n running from 5 to 9. Draw embeddings of each of these graphs in the appropriate compact surfaces.

   (b) Use the diagrams produced in (a) to help you to draw

      (i) a map on $2T$ needing 8 colours,

      (ii) a map on $3T$ needing 9 colours.

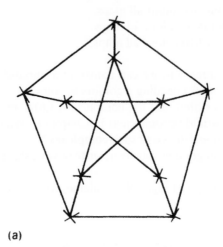

(a)

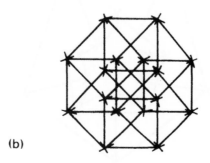

(b)

Fig. 6.9.2.

(c) Using Theorem 6.5.2, find $\gamma(K_{4,4})$ and $\gamma(K_{5,6})$. Draw embeddings of these graphs in the appropriate compact surfaces.

(12) Let C_n denote the graph formed by the vertices and edges of an n-dimensional cube.

(a) Prove that C_4 is non-planar.

(b) Find an inequality for $\gamma(C_n)$.

(c) Determine $\gamma(C_4)$.

(13) A circuit board is to be constructed to take circuits of the following type: there are to be three distinct sets of points, A, B, and C containing 1, m and n points respectively, and no point may be connected to a point in the same set.

Find a lower bound for the number of holes needed in the board to cope with all possible circuits of this type.

For the case $1 = m = 2$, $n = 3$, sketch a plane model of a board with the minimum number of holes needed to cope with all circuits, and on this plane model, draw a circuit which could not be placed on a board with less holes.

(14) (a) Embed K_6 in K and hence sketch a map on K needing exactly six colours.

     (b) Embed K_6 in P and hence sketch a map on P needing exactly six colours. (Compare Exercise 6.9.1.)

(15) Suppose that the graph G is embedded in the *non*-orientable surface M so that $\chi(M) = V - E + F$. If this embedding is such that $\chi(M)$ is a maximum, for such *non*-orientable embeddings, we say that we have a *minimal non*-orientable embedding (minimal with respect to what, this time?), and we say that $\chi(M)$ is the *non-orientable characteristic* of G, denoted $\beta(G)$.

     (a) Find inequalities for $\beta(G), \beta(K_n), \beta(K_{m,n})$ comparable with the inequalities for $\gamma(G)$, $\gamma(K_n)$, $\gamma(K_{m,n})$ found in section 6.5.

     (b) In Exercise (14) above, you have shown that your inequality for $\beta(K_n)$ gives equality if $n = 6$. Show that your inequality for $\beta(K_{m,n})$ also gives equality if $m = n = 3$.

     (c) Investigate the possibility of equality for other small values of n, or of m and n.

     (d) Prove that the bounds obtained for $\gamma(K_n)$ and $\beta(K_n)$ are equal for $n = 3 + 4m$ $(m = 1, 2, \ldots.)$ and for $n = 4 + 4m$ $(m = 0, 1, 2, \ldots.)$ *only*.

7

Vector fields on surfaces

7.1 A WATER PROOF

Consider a totally dry earth on which the landscape has been distorted slightly in places in order to exclude all plateaus, ridges, and level rims. In this new landscape, every mountain top reaches a peak at a single point and every valley reaches a pit (bottom) at a single point.

Fig. 7.1.1(a) shows the type of occurrence we are trying to exclude, while Fig. 7.1.1(b) illustrates how this could be done by means of a continuous distortion.

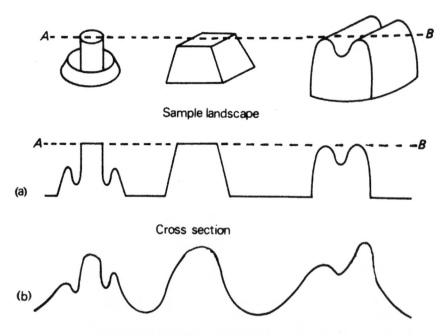

Fig. 7.1.1 — Continuous distortion shown in cross-section. Similar distortions take place in other directions.

In the new landscape, we call the peaks and pits **critical points**. We also attach this label to the **central** points of the **passes**.

By distributing hills and hollows around a point, we can construct many examples of critical points associated with passes. For example, Fig. 7.1.2 shows two instances called a **saddle point** and a **monkey saddle point**, respectively.

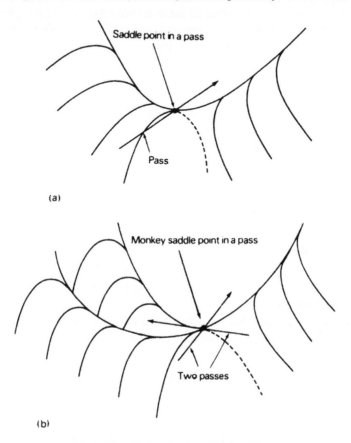

Fig. 7.1.2 — Passes showing saddle points.

A **pass** moves from a hollow, through the critical point, into the adjacent hollow. hence, the number of passes through the critical point is one less than the number of hollows adjacent to the critical point.

For convenience, let us further distort the landscape so that no two critical points lie at the same height above 'sea level'.

Now suppose that water gradually rises to cover the earth's surface, beginning at the level of the lowest pit and rising to cover the highest peak.

The lakes formed in the pits will eventually join together as they engulf the passes. Each pass, as it is covered in this way, either links two previously distinct lakes, or links a single lake with itself to form a moat around higher ground.

Suppose that P_1 passes eventually link distinct lakes, and that there were B pits in the original landscape. Eventually, the lakes formed in the B pits are all linked to form one large lake. Hence

(1) $B = P_1 + 1$.

Now suppose that P_2 passes eventually form moats. Each moat has two boundaries. When the water has risen so that all the inner boundaries have disappeared, there is still one last outer boundary surrounding 'Mount Everest'. This is the last to go. Hence, if there were M peaks originally, then

(2) $M = P_2 + 1$.

Adding (1) and (2), we obtain the mountaineer's equation:

(3)   *Peaks + Pits − Passes* $= 2 = \chi(S)$.

As we shall see later, this is a special case of a result which holds for any compact surface.

7.2   HAIRY SURFACES

Suppose that the compact surface M is covered with short, fine hairs. The problem is to brush the hairs flat in such a way that at *no* point on M is there a sudden reversal in the direction of the hairs.

In Fig. 7.2.1, we show two unsuccessful attempts to do this on a sphere S. By contrast, Fig. 7.2.2 shows two successful attempts to do this on a torus T.

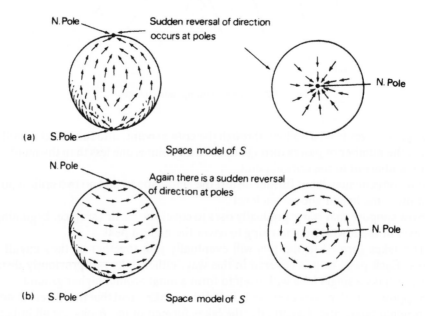

Fig. 7.2.1.

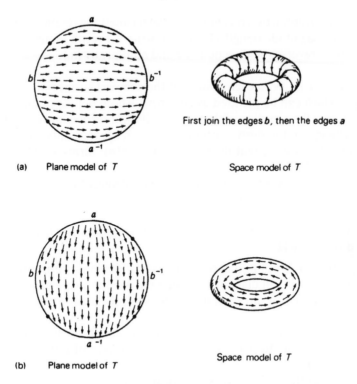

First join the edges *b*, then the edges *a*

(a)   Plane model of *T*   Space model of *T*

(b)   Plane model of *T*   Space model of *T*

Fig. 7.2.2.

To be more precise, let us make some definitions. First, we ask the reader to recall the elementary idea of a vector as a quantity having magnitude and direction. It is useful to picture a vector as represented by an arrow, whose length is proportional to the magnitude, and whose direction is that of the vector. In order to add two vectors, we assume that the representative arrows have their tails located at the same place, and then we use the parallelogram rule, as shown in Fig. 7.2.3.

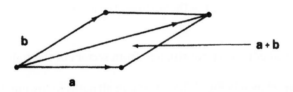

Fig. 7.2.3.

To multiply a vector **a** by a real number (scalar) m⩾0, we retain the direction of **a** but multiply its magnitude (length) by m. We write the result as **ma**.

For m < 0 we reverse the direction of **a** and multiply by |m|.

DEFINITION 7.2.1   With each point p of the compact surface M, we associate a vector, $v(p)$, which can be pictured as an arrow, whose tail is positioned at p, and whose direction is the direction of the vector **v**. If the direction and magnitude of the vector $v(p)$ change continuously as p varies, *except* at a *finite* number of points, called **critical points**, then we say that the set of vectors, $v(p)$, forms a **vector field** on the compact surface M.

It is sometimes convenient to distinguish a special kind of vector field in which the vectors have a fixed magnitude.

DEFINITION 7.2.2   A vector field, whose vectors $v(p)$ remain *fixed* in *magnitude*, is called a **direction field**.

In our problem of brushing hair to lie in certain directions, we are concerned with direction fields. In fact, all diagrams showing the general *shape* of vector fields are essentially illustrations of direction fields. We shall see many examples of these in this chapter.

In any vector field, the critical points must be isolated, because there are only a finite number of them. In particular, there are *no lines* of critical points as in a 'parting' of the hair.

Let us consider the various types of critical point.

The critical point shown in Fig. 7.2.4(a) is called a **centre** and the one shown in Fig. 7.2.4(b) is called a **focus**.

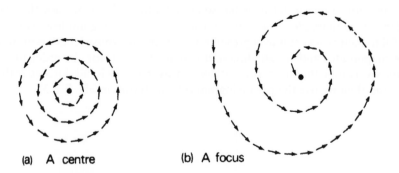

(a) A centre   (b) A focus

Fig. 7.2.4 — The two types of critical point.

All other critical points are constructed from **sectors** of the following types.

(1) **Elliptic sector**, shown in Fig. 7.2.5(a), where all paths following the arrows begin and end at the critical point.
(2) **Parabolic sector**, shown in Fig. 7. 2.5(b), where all paths lead away from the critical point *or* all paths lead towards the critical point.

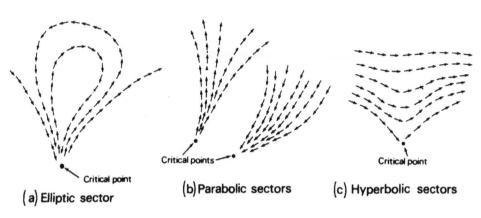

Critical points

Critical point

Critical point

(a) Elliptic sector   (b) Parabolic sectors   (c) Hyperbolic sectors

Fig. 7.2.5.

(3) **Hyperbolic sector**, shown in Fig. 7.2.5(c), where all paths sweep past the critical point.

The paths dividing each sector from the next at a critical point are called **separatrixes**.

Some examples of critical points constructed from sectors are shown in Fig. 7.2.6.

A critical point with just parabolic sectors is called a **node**. The node shown in Fig. 7.2.6(a) is called a **sink** and that in Fig. 7.2.6(b) is called a **source**.

A critical point with just elliptic sectors is called a **rose**. The particular kind of rose shown in Fig. 7.2.6(c) is called a **dipole**.

A critical point with just hyperbolic sectors is called a **cross point**. A **saddle point** is a particular example of a cross point.

With each critical point p we associate a number called the **index** of p and evaluated as follows.

Describe a circle C around p, so that *no other* critical point lies inside or on C. Imagine a point x starting at some point q on C and travelling in an anti-clockwise sense along the circumference of C back to q. At each position of x imagine an arrow drawn in the direction of the vector field at that point.

DEFINITION 7.2.3   Let p be a critical point in a vector field on the compact surface M. The index of p is the number of anti-clockwise revolutions made by the arrows, described above, as x travels from q along the circumference of the circle C, in an anti-clockwise sense.

This process is illustrated in Fig. 7.2.7 for a focus, whose index is found to be 1.

A similar diagram shows that the index of a centre is also 1.

Notice that the 'direction' taken by the focus or the centre, i.e. clockwise or anti-clockwise, does *not* affect the value of the index of the critical point.

If p is any other critical point, then the elliptic, parabolic, and hyperbolic sectors contribute to the index of p as follows.

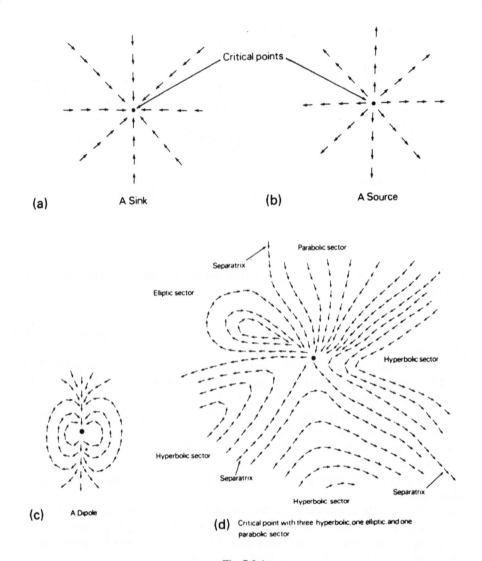

Fig. 7.2.6.

Draw the circle C with centre p. Join p to the points at which the separatrixes meet the circle C. This divides the circle into sectors in the ordinary geometric sense. Let $\alpha_1, \ldots, \alpha_a$ be the angles of those sectors of the circle associated with the pairs of separatrixes belonging to the elliptic sectors. Similarly let $\beta_1, \ldots, \beta_b$ be the corrresponding angles for the parabolic sectors, and $\gamma_1, \ldots, \gamma_c$ for the hyperbolic sectors.

From Fig. 7.2.8(a), we can see that when the circle C passes through an elliptic sector with associated angle α_i, the arrows on C can *be taken* to turn through $\alpha_i + \pi$ in an anti-clockwise sense; see *note* below.

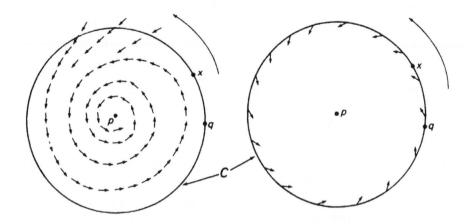

Fig. 7.2.7 — A focus at p, index 1. The arrows on C turn through one complete anti-clockwise revolution as x travels around the circle C.

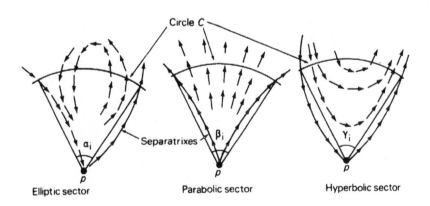

Fig. 7.2.8.

Note that we are calculating the angle of turn for each sector as though the flow was along the radii drawn from the point p to the points of intersection of the separatrixes with the circle C. Although this may not give the correct angle of turn for individual sectors, it does give the correct angle of turn for C as a whole, since the 'turn' lost by one sector is gained by the neighbouring sector.

Fig. 7.2.8(b) and (c) show that the corresponding angles for the parabolic and hyperbolic sectors are β_i and $\gamma_i - \pi$, respectively, in an anti-clockwise sense.

Thus, in travelling once round C in an anti-clockwise sense, the arrows on C turn anti-clockwise through

$$(\alpha^1 + \pi + \ldots + (\alpha_a + \pi) + \beta_1 + \ldots + \beta_b + (\gamma_1 - \pi) + \ldots + (\gamma_c - \pi)$$
$$= (\alpha_1 + \ldots + \alpha_a) + (\beta_1 + \ldots + \beta_b) + (\gamma_1 + \ldots + \gamma_c) + a\pi - c\pi .$$

But

$$(\alpha_1 + \ldots + \alpha_a) + (\beta_1 + \ldots + \beta_b) + (\gamma_1 + \ldots + \gamma_c) = 2\pi .$$

Thus we have established:

FORMULA 7.2.1 Index of $p = 1 + (a - c)/2$, where a = the number of elliptic sectors and c = the number of hyperbolic sectors.

Notice that the parabolic sectors do not contribute to the index of p.

We are now in a position to evaluate the index of *any* critical point on a compact surface.

This means that, when any vector field on the compact surface M, we can associate a number, namely the sum of the indices of the critical points of the vector field. If we make the fairly big *assumption* that this number *depends only* on the compact surface M, then we can relate it to $\chi(M)$, when M is orientable, as follows.

Let C be a complex on the orientable surface M with V vertices, E edges, and F faces. We construct a vector field on M by placing a source at each vertex, a sink within each face, and a saddle point within each edge, as shown in Fig. 7.2.9. This *can* be done for *any orientable* compact surface.

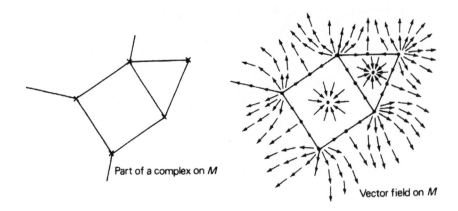

Part of a complex on M

Vector field on M

Fig. 7.2.9.

The complete vector field then has V sources, each with index 1, E saddle points, each with index -1, and F sinks, each with index 1.

Thus the sum of these indices is

$$V - E + F = \chi(M) .$$

Since we are assuming that this sum depends only on M, it follows that the sum of the indices is equal to $\chi(M)$ for *any* vector field on M.

This result is often called the index theorem, which we state formally as

THEOREM 7.2.1 (The index theorem)   The sum of the indices of the critical points of any vector field on the *orientable* compact surface M is $\chi(M)$.

Applying this theorem to the sphere S, for which $\chi(S) = 2$, we obtain the 'hairy-ball' theorem, perhaps conjectured by hairdressers for hundreds of years, before being finally proved by Poincaré and Brouwer!

THEOREM 7.2.2 (The hairy-ball theorem)   Every vector field on a sphere has a critical point.

Notice that Theorem 7.2.1 tells us that if a vector field exists on an orientable compact surface, and we introduce a *new* critical point in the vector field, then *other* critical points *must appear* elsewhere in the vector field in order to keep the sum of the indices equal to $\chi(M)$. For example, Fig.7.2.10 shows how a cross-point, with index -1, can appear on the 'hairy-torus' of Fig. 7.2.2(b), when, by combing the hair, we try to introduce a centre with index 1.

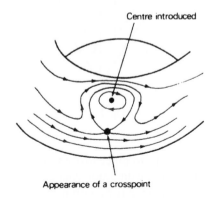

Centre introduced

Appearance of a crosspoint

Fig. 7.2.10.

7.3 INTERPRETATIONS OF THE INDEX THEOREM

In everyday life, we meet many situations where Theorem 7.2.1 gives us interesting information. Weather maps, contour maps, diagrams of streamlines, all describe part of what could be a vector field on a sphere S.

If we draw contour lines on a piece of sculpture, or simply consider the lines of the grain in a wood sculpture, we can easily find examples of vector fields, or, more specifically, direction fields, on *any orientable* compact surface. Remember, though, that the lines in a vector field have direction and that change of direction must occur smoothly except at a finite number of isolated points, i.e. the critical points.

Fig. 7.3.1 shows that it is *not* always possible to place arrows on a contour map so as to get a direction field. For example, if we impose arrows in an anti-clockwise sense on the triangles in region *A*, this determines a clockwise direction for the triangles in region *B*, which in turn determines an anti-clockwise direction for the triangles in region *C*. But the latter then produces a line of points *l*, along which the direction changes suddenly.

Line *l* of points along which the direction changes suddenly

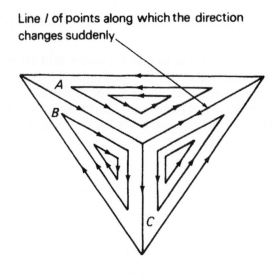

Fig. 7.3.1.

In a similar way, *any* attempt to place arrows on these lines *always* produces a complete line of points at which the direction changes suddenly. To remedy this, we must change the map, as we changed the contours in the water proof of section 7.1, by continous distortion, so that all critical points become isolated. This involves introducing more contour lines in order to produce hollows between the hills.

Once we have obtained a vector field on an orientable compact surface, we can use Theorem 7.2.1 to obtain information about a portion of the vector field by considering the 'inside' of a closed loop in the vector field. This technique is illustrated in the following sections.

7.4 LAKES

Consider the flow of water on the surface of a lake. To avoid the disturbances occurring at the edge, we take the boundary of the lake at a certain distance from the shore and suppose that the water flows around this boundary line.

In section 7.2, we consider the various types of critical point which can occur. But can each of these occur in practice on the surface of the lake? As we shall see, the answer to this question is a *qualified* yes!

It is easy to see how we could try to visualize the introduction of a source or a sink in the flow. We could simply place a small tube in the flow and either pump water in or pump water out through the tube.

To introduce a vortex into the flow, with its attendant central critical point, we could place a small cylinder in the water and rotate it. The critical point could then be imagined at the centre of the cylinder.

Combining the two techniques, we can introduce the effect of any one of the different types of critical point into the flow. For example, by introducing both a source and a sink, as above, and then viewing the flow from a distance, we obtain the effect of a dipole, as shown in Fig. 7.4.1.

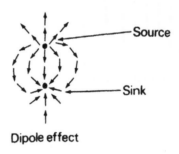

Dipole effect

Fig. 7.4.1.

The closer together the sink and source are placed, the more accurately does the configuration approximate a dipole.

The more complicated critical point shown in Fig. 7.4.2(a) could be induced into the flow by the configuration shown in Fig. 7.4.2(b), where we have four rotating cylinders and a pumping tube, punctured on one side.

In general, we can see that, provided we view the flow from a great distance, part of any vector field possible on a sphere S, can be constructed on the surface of a lake.

However, in nature, certain critical points are more common than others. In practice, unless we happen to be on the equator, every time we try to produce a source or a sink by pumping water through a tube, the rotation of the earth, in fact, conspires with the source or sink to produce a focus. It would appear that sources and sinks occur rarely in nature in fluid flow. On the other hand, foci and centres often appear naturally. Think of the flow past a bridge support or the flow induced by dipping an oar in the water. Taken in conjunction with the index theorem this implies that cross points are also very common.

7.5 ISLANDS IN LAKES

Other critical points, while less common in a natural setting, are frequently considered by mathematicians, particularly when modelling a real situation. For

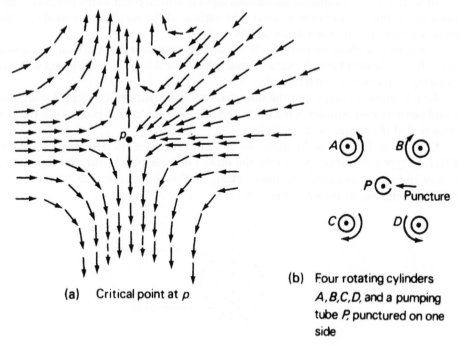

(a)   Critical point at p

(b)   Four rotating cylinders
      A,B,C,D, and a pumping
      tube P, punctured on one
      side

Fig. 7.4.2.

example, in considering a uniform flow past an island in the lake, a mathematician would pretend that the effect of the island on the flow could be modelled by replacing the island by a dipole placed at the centre of the island, as shown in Fig. 7.5.1.

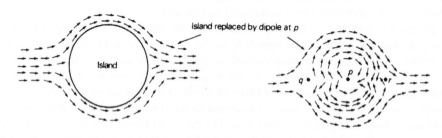

Fig. 7.5.1.

The flow of the dipole inside the region covered by the island has, of course, no counterpart in the real situation of island and lake.

Notice that, as well as the dipole centred at p, we also have cross points at q and r. The total index for this configuration is $-1 + 2 - 1 = 0$.

The alternative to uniform flow past an island is circulation round an island, as shown in Fig. 7.5.2.

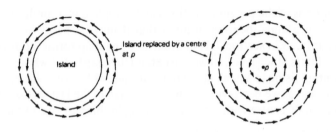

Fig. 7.5.2.

This time a mathematician would pretend that the effect on the flow was caused by a centre, index 1, placed at the centre of the island. Of course, throughout this work, we are idealizing the situation by taking islands to have circular boundaries.

To extend the flow in the lake to a flow over the whole of a sphere S, we treat the whole of the outside boundary of the lake as the boundary of a single large island with water circulating round it. Thus the rest of S, outside the lake, is replaced by a centre with index 1.

Now suppose that water flows uniformly past I_1 islands, and circulates around I_2 islands. Then, replacing all these islands, and the outside of the lake, by dipoles or centres, as above, we obtain the following sum of indices for this vector field on S:

$$N = 0I_1 + I_2 + 1 \ ,$$

which equals $\chi(S) = 2$, by the Index Theorem 7.2.1.

Now the sum of the indices of the critical points *not actually in the lake* is:

$$2I_1 + I_2 + 1 \ .$$

Thus the sum of the indices of the critical points *in the lake* is:

$$N - (2I_1 + I_2 + 1) = 2 - (2I_1 + I_2 + 1)$$
$$= 1 - 2I_1 - I_2 \ .$$

We state this result as

COROLLARY 7.5.1   If the water in a lake flows uniformly past I_1 islands, and circulates around I_2 islands, then the sum of the indices of the critical points of the flow *in the lake* is:

$$1 - 2I_1 - I_2 \ .$$

7.6 ISLANDS

The considerations of the previous section can be applied in a similar way to discuss contour maps of islands, bearing in mind that it may be necessary to distort the landscape slightly, as in section 7.1, in order that the map forms a direction field.

The shore line is at sea level, so we can extend the map to the whole of S by having a single mountain rising to a peak at the opposite side of S to the island. We deal with lakes on the island as we dealt with islands in the lake in the last section. Each lake is replaced by a single centre with index 1. Arguing in this way, we obtain:

COROLLARY 7.6.1   On an island

$$Peaks - Passes + Pits + Lakes = 1 \ .$$

The reader is invited to fill in the details.

7.7 VECTOR FIELDS AND DIFFERENTIAL EQUATIONS

We can relate a vector field on a subset of the plane, $\mathbb{R}^2$, to a pair of differential equations, as follows.

Suppose that a vector field is defined on some region D of $\mathbb{R}^2$, say a disc. Let the components of the vector at the point (x,y) in the vector field be $F(x,y)$ and $G(x,y)$.

With each point (x,y) in D, we associate a direction, represented by an arrow. In other words, we define a **direction field** on D, in accordance with Definition 7.2.2. Let this direction at (x,y) in D be the *direction* of the vector from the origin $(0,0)$ to the point $V(x,y) = (F(x,y),\ G(x,y))$, i.e. the point whose coordinates are $F(x,y)$ and $G(x,y)$. The situation is illustrated in Fig. 7.7.1.

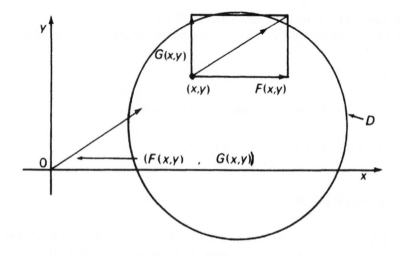

Fig. 7.7.1.

Thus the direction field can be obtained from the vector field by taking each vector to have magnitude 1. If we think of the typical point (x,y) in the vector field following a path in the direction specified by the direction field at the point (x,y), then the position of the point (x,y) depends on time, and we may write

(1) $dx/dt = F(x,y)$, $dy/dt = G(x,y)$.

In this way a vector field, defined on a set of points in the plane, determines a pair of differential equations.

Conversely, a pair of differential equations of the form (1) determines an equation $V(x,y) = (F(x,y), G(x,y))$, which, in turn, determines a vector field on some set of points in the plane.

If we are given a vector field in the form $V(x,y) = (F(x,y), G(x,y))$, then it is a simple matter to find the critical points (singularities) and sketch the paths determined by the corresponding direction field. We notice that the curves $F(x,y) = 0$ describe those sets of points (x,y) for which $dx/dt = 0$. In other words, they describe all points (x,y) at which the directions in the vector field are vertical. Similarly, the curves $G(x,y) = 0$ describe all points (x,y) at which the directions in the vector field are horizontal.

These curves will intersect at the critical points. In any *one* region *between* these curves, since no directions are horizontal or vertical and any change of direction is continuous, all directions must point towards just one quadrant. This information enables us to sketch the *shape* of the vector field.

EXAMPLE 7.7.1   Sketch the vector field

$$V(x,y) = (2xy, y^2 - x^2 - 1) .$$

Solution. The set of points in the vector field at which the directions are vertical is given by $2xy = 0$. This determines the lines $x = 0$ and $y = 0$.

If $y^2 - x^2 - 1 \geq 0$ for such a point, then the direction is upwards, and, if $y^2 - x^2 - 1 \leq 0$, then the direction is downwards.

Similarly, the set of points at which the directions are horizontal is given by $y^2 - x^2 - 1 = 0$. This determines the curves $y = \sqrt{(x^2 + 1)}$ and $y = -\sqrt{(x^2 + 1)}$.

Sketching all these curves, with appropriate directions indicated, gives Fig. 7.7.2(a)

Filling in the general shape of the vector field between these curves, gives Fig. 7.7.2(b).

EXAMPLE 7.7.2   *The Voltera prey–predator equations*

Suppose that we have two kinds of animals, one kind preying on the other. Let $x(t)$ be the size of the prey population and let $y(t)$ be the size of the predator population, both at time t.

Suppose that each population has a natural growth rate proportional to its size, say, ax and $-by$, respectively, where the difference in sign reflects the fact that an

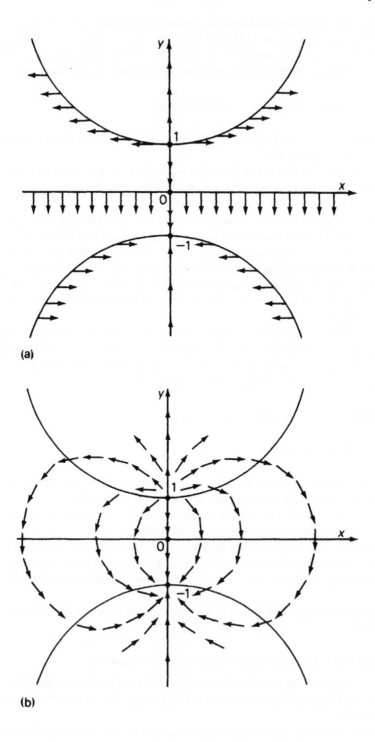

(a)

(b)

Fig. 7.7.2.

increase in one population is accompanied by a decrease in the other. We suppose that xy is a reasonable measure of the number of encounters between prey and predator. Such encounters will adversely affect the growth rate of the prey population, but will improve that for the predator population.

Combining this effect with the natural growth rate mentioned earlier, we obtain the equations:

$$dx/dt = ax - cxy \ , \qquad dy/dt = -by + dxy \ ,$$

where a, b, c, and d, are suitable non-negative constants.

These equations determine a vector field $V(x, y) = (ax - cxy, \ -by + dxy)$, which is sketched in Fig. 7.7.3.

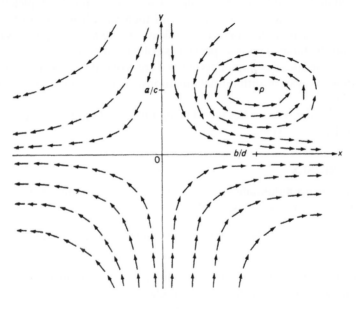

Fig. 7.7.3.

Naturally only the first quadrant, where the size of each population is a non-negative number, has any bearing on the real situation.

The flow around the centre at $p = (b/d, \ a/c)$ shows how the sizes of the two populations rise and fall periodically.

Finally, we remark that, while we have considered differential equations and vector fields only on the plane $\mathbb{R}^2$, or on subsets of the plane, consideration of these topics on general orientable compact surfaces is by no means just a mathematical curiosity. In fact, vector fields on the torus arise naturally from doubly periodic solutions of differential equations on the plane $\mathbb{R}^2$.

7.8 COMMENTS

In keeping with the general style of this book, we have kept our discussion of vector fields on surfaces, usually orientable and compact, at an elementary level. This approach may give the impression that the study of these topics is an interesting pastime without important applications. Such an impression is far from the truth.

Vector fields represent a link between analysis and combinatorial topology which goes back to the very beginnings of the subject and finds many important applications, both to other branches of mathematics and to the physical sciences.

Poincaré's pioneering work in topology, around 1895, grew out of his interest in differential equations and led to important contributions to the qualitative theory of these equations. In particular, his work threw light on the form of integral curves and the nature of critical points.

Brouwer, too, dealt with critical points of vector fields in 1911. In fact, this area of work is closely related to the fixed point theorems. The latter have not been considered in this book. To obtain some idea of what is involved, pick up a flat sheet of paper from the table, screw it up, and then squash it flat *inside* the region it originally occupied on the table. Then at least one point in the paper is in its original position on the table. This is a special case of a fixed point theorem.

Many of the later developments of large areas of topology were directly related to applications by way of differential equations, and this trend has continued to the present day.

For the reader who wishes to follow up the ideas of this chapter, we recommend Henle [16] as the next stage.

7.9   EXERCISES

(1) Write down a mountaineer's equation for a planet homeomorphic tó a torus T. What are your 'contours' based on?

(2) Show that there is always at least one point on the surface of the earth at which the wind is still.

(3) For each of the critical points shown in Fig. 7.9.1, find the index using Definition 7.2.3. Check your answers by using formula 7.2.1.

(4) Draw contour lines describing a monkey saddle point, and determine its index.

(5) Prove that any critical point with an odd number of sectors contains at least one parabolic sector.

(6) Among the critical points with no more than six sectors, find and sketch the distinct ones with index three.

(7) The fact that the direction of movement of the hands of a clock and the direction of movement of water down a sink are related can be explained by the rotation of the earth. How?

(8) Draw streamlines for lakes with the following features:
   (a) a spring (source), a sink, and two stagnation points,
   (b) as in (a), but with a boat at anchor,
   (c) two islands, a spring, and two sinks.

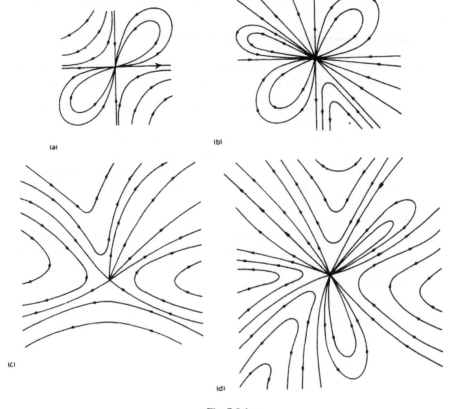

(a)   (b)

(c)

(d)

Fig. 7.9.1.

(9) Sketch the following vector fields and determine the index of each critical point.
   (a) $V(x,y) = 1,2)$
   (b) $V(x,y) = (-y,x)$
   (c) $V(x,y) = (1 - x^2, 2xy)$
   (d) $V(x,y) = (x, x^2 - y)$
   (e) $dx/dt = y$, $dy/dt = 1 - x^2$
   (f) $dxd/t = y^2 - x$, $dy/dt = xy - y$
   (g) $dx/dt = x(y^2 - 1)$, $dy/dt = -y(x^2 - 1)$
   (h) $dx/dt = x^2 - 3xy^2$, $dy/dt = y^3 - 2xy$.

(10) In some cases the paths followed by the points in a vector field (the integral curves of the associated differential equations) can be determined by solving the single differential equation

$$dy/dx = G(x,y)/F(x,y)$$

   (a) Find the equations describing the integral curves of $V(x,y)) = (3x,y)$ and sketch the paths.
   (b) Find the equations describing the integral curves of the prey–predator vector field.

(11) Contour lines can be related to vector fields by way of the function $\phi(x,y)$, where

$$V(x,y) = (\partial\phi/\partial y(x,y), \ -\partial\phi/\partial x(x,y))$$

Then the curves $\phi(x,y) = $ constant can represent curves of constant height, potential, etc. Find the function ϕ for each of the vector fields in Exercise 9, above.

(12) The **dual** of the vector field $V(x,y) = (F(x,y), \ G(x,y))$ is the vector field $V^*(x,y) = (-G(x,y), \ F(x,y))$.

    (a) How is the direction of the arrow at the point p in $V(x,y)$ related to the direction of the arrow at the same point p in $V^*(x,y)$?

    (b) The integral curves of $V^*(x,y)$ are called **orthogonal trajectories** of the integral curves of $V(x,y)$. Sketch the vector fields $V^*(x,y)$ for each of the examples in Exercise 9, above.

    (c) How will $V^{**}(x,y)$ be related to $V(x,y)$?

    (d) If the curves $\phi(x,y) = $ constant represent curves of constant height (pressure, etc.) for the vector field $V(x,y) = (\partial\phi/\partial y(x,y), \ -\partial\phi/\partial x(x,y))$, then the *same* curves for the **dual** vector field $V^* = (\partial\phi/\partial x, \ \partial\phi/\partial y) = \nabla\phi$, the **gradient** of ϕ, represent what?

(13) For each of the vector fields on the compact surfaces shown in Fig. 7.9.2, mark the critical points, determine their indexes, and verify the conclusion of Theorem 7.2.1.

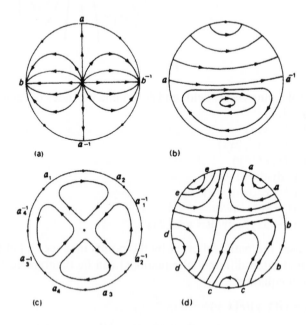

Fig. 7.9.2.

(*Notice* that, although one of these surfaces is non-orientable, the sum of the indices is still equal to the Euler characteristic in this case.)

8

Plane tessellation representations of compact surfaces

8.1 PLANE EUCLIDEAN GEOMETRY

In this book we have already assumed on several occasions that the reader has some familiarity with the 'ordinary' Euclidean geometry of everyday life. This is the geometry which describes, with a good degree of accuracy, the space in which we live, provided our distances are small enough. For example, since we live on the surface of the earth, which is very nearly a sphere, the shortest distance between two points is not a straight line in the Euclidean sense but an arc of a 'great circle', i.e. a circle on the sphere whose radius is the same as that of the sphere. But as long as we are dealing with distances of less than a mile, say, the error in assuming that the arc of the great circle is a straight line is negligible.

Some 2500 years ago, the ancient Greeks made considerable contributions to the study of geometry. Most of their results were collected together by Euclid, who presented then in the form of axioms and deductions therefrom (theorems) that has become so familiar and for centuries has been a model of how mathematics should be presented. Unfortunately, his axioms are not complete. For example, he assumes, but never states as an axiom, that lines may be extended indefinitely. In fact, Hilbert, in the 1930 edition of his *Grundlagen der Geometrie*, needed some 20 axioms, as compared to Euclid's 5 axioms, in order to characterize Euclidean plane geometry precisely. Euclid's 5 axioms are:

(1) It is possible to draw a straight line from any point to any point.
(2) It is possible to produce a finite straight line continuously in a straight line.
(3) It is possible to describe a circle with any centre and radius.
(4) All right angles are equal to one another.
(5) If a straight line meets two straight lines so that the interior angles on the same side are less than two right angles, then the two straight lines, if produced indefinitely, meet on that side on which the angles are less than two right angles.

In the exercises we illustrate how these 5 axioms are not sufficient.

A convenient way of handling Euclidean geometry, which avoids the tedium of using a large set of axioms, is to use a concrete example, or *model*, of the geometry. One of the best known models makes explicit use of the concept of *distance* between two points. We take two mutually perpendicular straight lines as axes, and, with equal scales along the two axes, we define the position of a point by a pair of coordinates (x, y). The distance between two points (x_1, y_1) and (x_2, y_2) is:

$$\sqrt{(x_1 - x_2)^2 + (y_1 - y_2)^2},$$

where the coordinates are certain real numbers. This gives the usual Euclidean plane coordinate geometry already used in this book.

Curves, such as a circle of radius r centred at (x_0, y_0), or an ellipse centred at the origin with semi-axes a and b, are described by equations.

$$(x - x_0)^2 + (y - y_0)^2 = r^2 \quad \text{and} \quad x^2/a^2 + y^2/b^2 = 1,$$

respectively. Fig. 8.1.2 illustrates these ideas.

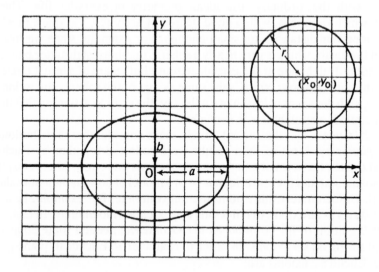

Fig. 8.1.2.

Just as, in Chapter 1, we defined topology, or 'rubber-sheet' geometry, to be the study of those properties of space which are unchanged by continuous transformations, so we can consider Euclidean geometry to be the study of those properties of space which are unchanged by transformations leaving fixed the distance between any two points. Such a transformation is called an (Euclidean) **isometry**.

As far as *plane* Euclidean geometry is concerned, isometries are precisely those transformations obtained by successive applications of *three basic* types of transformation, namely **rotations** about a point, **reflections** in a line, and **translations**: see Fig. 8.1.3. In fact, since a rotation about a point O can be obtained by reflections in two lines intersecting at O, and a translation can be obtained by reflections in parallel lines, all isometries of the plane can be obtained by successive reflections, see Coxeter [5] and Lockwood & Macmillan [21].

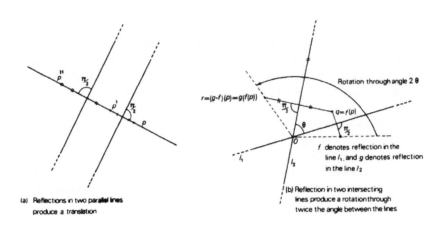

Fig. 8.1.3.

The action of the isometry f on point p is usually written $f(p)$. Thus f moves the point p to the position previously occupied by $f(p) = q$.

The reader should have no trouble in checking.

THEOREM 8.1.1

(1) There is an isometry of the plane leaving everything fixed. It is called the **identity** transformation.
(2) The product of two isometries f and g, written $g \circ f$, meaning the isometry f followed by the isometry g, is also an isometry.
(3) To each isometry f there is a **unique inverse** isometry, written f^{-1}, which 'undoes' the effect of f, and transforms the plane back to its original configuration before the application of f.
(4) The product of isometries satisfies the associative rule, i.e. if f, g and h are isometries, then $h \circ (g \circ f) = (h \circ g) \circ f$.

Since the Euclidean plane isometries define Euclidean plane geometry, we would expect to be able to prove the theorems of this geometry by the direct use of isometries; see Jeger [17]. As an example of this, we prove

The angles at the base of an isosceles triangle are equal.

PROOF Consider Fig. 8.1.4.

Here we have an isosceles triangle ABC with $AB = AC$. Draw the line AD perpendicular to BC. Let f be the reflection of the plane in the line AD. (This line, of course, extends indefinitely in both directions.) Because $AB = AC$, and f preserves the distance between any two points, $f(C)$ must be B and BD must equal DC. Thus, in particular, the angle $D\hat{C}A$ must fit precisely over the angle $A\hat{B}D$ under the isometry f. Hence $A\hat{B}C = A\hat{C}B$.

A set of transformations satisfying conditions (1)–(4) of Theorem 8.1.1 is an example of one of the most important concepts in modern mathematics, namely the concept of a *group*. In this case, we have the group of isometries of the Euclidean plane, often called the Euclidean group in 2 dimensions, $\mathbf{E}_2$.

Euclidean plane geometry is the study of those properties of space which are unchanged under the action of any of the elements of $\mathbf{E}_2$. The fact that *any* Euclidean plane isometry is a product of reflections is expressed by saying that E_2 is *generated* by reflections.

Similar considerations apply to n-dimensional Euclidean space $\mathbb{R}^n$ and the associated Euclidean group $\mathbf{E}_n$; see Coxeter [5].

In the rest of this chapter, we shall be concerned not only with Euclidean plane geometry but also with *non*-Euclidean plane geometry. But before considering the latter and how it arose, it will be useful for later work to look more closely at the idea of a group.

8.2 GROUPS

Although the idea of symmetry has played an implicit part in mathematics and art since the earliest times, only since around 1830 has a conscious and explicit study of symmetry and its role in human understanding been made. Apart from the insight that this study has given in most branches of mathematics, it has also been invaluable in many physical sciences, such as physics and chemistry, in particular, elementary particle physics and molecular structure.

In order to make a systematic study of symmetry, mathematicians introduced the concept of a group defined as follows.

DEFINITION 8.2.1   A **group** G is a set of objects denoted by a, b, c, $\ldots$, say, together with a '**multiplication**' defined on the set such that the '**product**' of any two elements a and b of the set is a *unique* element ab of the set. The following axioms must be satisfied.

(1) $(ab)c = a(bc)$, the associative rule,
(2) there exists an element e of the set with the property that $ae = ea$ for *all a* in the set,
(3) with each element a in the set is associated an element a^{-1} with the property that $aa^{-1} = a^{-1}a = e$.

If, in addition, we have

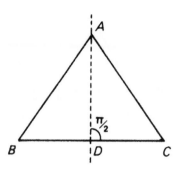

Fig. 8.1.4.

(4) $ab = ba$, for *all* elements a, b, then the group is said to be **abelian** or **commutative**.

Note that the 'e' mentioned in condition (2) is *unique*. It is called the **identity** of the group. Similarly, the element a^{-1} is *unique* for a particular a. It is called the **inverse** of a.

The following proofs of these statements provide examples of how one argues from the definition of a group.

Uniqueness of e
If e' also satisfies $ae' = e'a = a$ for *all* a in the group, then, in particular, $ee' = e'e = e$. But also $ae = ea = a$ for *all* a, hence $ee' = e'e = e'$. Thus $e = e'$. Hence e is unique.

Uniqueness of a^{-1}
Suppose also that $aa' = a'a = e$, for some a'. Then $a^{-1} = a^{-1}e = a^{-1}(aa') = (a^{-1}a)a' = ea' = a'$, using the associative rule. Thus $a^{-1} = a'$, which shows that a^{-1} is unique.

We usually write $aaa \ldots aa$ (n factors) as a^n, as in 'ordinary' algebra. Then $a^{-1}a^{-1} \ldots a^{-1}a^{-1}$ (n factors) is written a^{-n}. We take a^0 to be e. The usual laws of indices are obeyed.

In order to 'measure' the symmetry of an object in space, we consider the group of transformations which leave the object *apparently fixed* in space. For example, there are essentially six distinct transformations (symmetries) which leave an equilateral triangle apparently fixed in space. These are shown in Fig. 8.2.1. They form a group under the usual product of transformations. This group is called the **symmetry group** of the equilateral triangle.

It is straightforward to check that the set of Euclidean isometries introduced in section 8.1 is a group, where the product $g{\circ}f$ is the isometry obtained by following the isometry f with the isometry g. In fact, this follows at once from Theorem 8.1.1. Other simple examples of groups are as follows: first the set of integers $\mathbf{Z} = \{\ldots -1, 0, 1, 2, \ldots\}$ with the ordinary addition of integers as the group multiplication, the identity is 0 and $-a$ is the inverse of a; next the set $C = \{1, -1, i, -i\}$. This is a group under the usual multiplication of complex numbers, where $i^2 = -1$. Both these

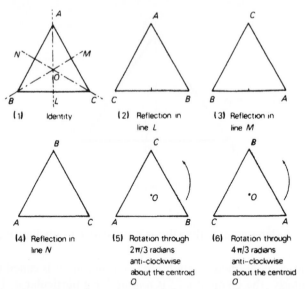

(1)   Identity

(2)   Reflection in line L

(3)   Reflection in line M

(4)   Reflection in line N

(5)   Rotation through $2\pi/3$ radians anti-clockwise about the centroid O

(6)   Rotation through $4\pi/3$ radians anti-clockwise about the centroid O

Note that each transformation is carried out on the triangle in the starting position shown in (1).

Fig. 8.2.1.

groups are examples of cyclic groups, i.e. groups generated by a single element. **Z** is generated by 1 and C is generated by i. In the latter case we have $C = \{i, i^2, i^3, i^4\}$.

The theory of groups has grown over the last 150 years to be one of the most important and extensive branches of mathematics, with ramifications in almost all other branches of mathematics. In this section, our modest task is to describe a few basic properties of groups which are of use in the rest of this book. Those readers who would like to pursue the topic further are referred to Weyl [33] and Gardiner [10] for the basic ideas, and to Macdonald [23], Rotman [27], and Johnson [18] for those aspects of the subject of particular importance in topology.

In section 8.1, we mentioned that the Euclidean group $\mathbf{E}_2$ could be generated by reflections, i.e. every element of $\mathbf{E}_2$ is a product of a *finite* number of reflections. It is this idea of a group generated by certain of its elements which proves to be of particular interest in topology.

First we develop the idea of a *free group*. Take a set X, which, for simplicity, we assume is finite, although the theory can be developed for any set, finite or infinite.

Let $X = \{x_1, x_2, x_3, \ldots, x_n\}$. Take another set, X^{-1}, in one–one correspondence with X, and label its elements x_i^{-1}, i = 1, 2, 3, ..., n. Thus

$$X^{-1} = \{x_1^{-1}, x_2^{-1}, x_3^{-1}, \ldots, x_n^{-1}\}.$$

This notation is chosen because of its convenience later. You could, if you wish, denote X^{-1} by $\{x_1', x_2', \ldots, x_n'\}$, say.

Now let

$$\overline{X} = X \cup X^{-1} = \{x_1, \ldots, x_n, x_1^{-1}, \ldots, x_n^{-1}\} \ .$$

DEFINITION 8.2.2.   Any expression of the form $y_1 y_2 y_3 \ldots y_t$, where $y_i \in \overline{X}$, $i = 1, 2, 3, \ldots, t \geqslant 0$, is called a **word** in the elements of $\overline{X}$. The **length** of a *word* is the integer t. A *word* of length 0 is written e.

EXAMPLE 8.2.1   Let $X = \{x_1, x_2, x_3\}$. *Then some typical words* in

$$\overline{X} = \{x_1, x_2, x_3, x_1^{-1}, x_2^{-1}, x_3^{-1}\}$$

are

$$x_2^{-1}, \ x_1^{-1} x_2 x_3 x_2^{-1}, \ x_1 x_3 x_2 x_1 x_2 x_3^{-1}, \ x_1 x_2 x_1^{-1} x_2^{-1}$$
$$x_1^{-1} x_1 x_2^{-1} x_2^{-1}, \ x_2 x_2 x_2 x_1^{-1} x_3^{-1} x_2 x_3, \ e.$$

Note that two *words* are *equal* if and only if they have the same length and corresponding letters are identical.

DEFINITION 8.2.3   The **product** uv of the two words $u = y_1 y_2 y_3 \ldots y_t$ and $v = y_1' y_2' \ldots y_s'$, *neither* equal to e, is the word

$$uv = y_1 y_2 y_3 \ldots y_t y_1' y_2' y_3' \ldots y_s'.$$

We define the products eu and ue to be u.

Notice the obvious similarity between the *words* defined above and the *words* introduced in section 3.2. This is no accident. The connection will be explained in Chapter 10. For the moment let us continue with the construction of free groups.

DEFINITION 8.2.4   Two *words* are said to be **equivalent** if they differ only in the presence of pairs of symbols of the form $x_i x_i^{-1}$ or $x_i^{-1} x_i$.

EXAMPLE 8.2.2 $x_1^{-1} x_1 x_2 x_3 x_3^{-1} x_4^{-1} x_4^{-1}$, $x_2 x_4^{-1} x_4^{-1}$; and $x_1^{-1} x_1 x_2 x_4^{-1} x_4^{-1}$ are all equivalent *words*.

    The reader should compare this idea with the idea of equivalence of rational numbers (fractions). Thus 2/3, 4/6, 12/18, and $-8/-12$, are all equivalent fractions. We could denote the set of all these equivalent fractions by [2/3], say, if we wanted to be precise in our notation. In practice, we gloss over this question, and talk about 2/3, 12/18, and so on, to mean the *same* rational number.
    In a similar way, by $[y_1 y_2 y_3 \ldots y_t]$ we mean the set of *all words equivalent to the word* $y_1 y_2 \ldots y_t$. In practice, we use *any* of the *words* in this set as a *representative* of the set.
    Let F be the set of all expressions of the form $[y_1 y_2 \ldots y_t]$. Then

$$[y_1 y_2 \ldots y_t] = [y_1' y_2' \ldots y_s']$$

if and only if $y_1 y_2 \ldots y_t$ is *eqivalent* to $y_1' y_2' y_3' \ldots y_s'$.

DEFINITION 8.2.5   We define the **product** $[u][v]$ to be $[uv]$, where uv is the product of the two *words* u and v given in Definition 8.2.3.

Note.   It should be checked that this definition does not depend on the choice of the representatives u and v. This is a technical matter that we avoid here, see for example, Macdonald [23]. In fact, the whole of this development turns on the idea of an equivalence relation on a set and the associated equivalence classes; in our case the expressions $[y_1 y_2 y_3 \ldots y_t]$. For details the reader is referred to Gardiner [11].

We are now able to state

THEOREM 8.2.1   With the product of two elements $[u]$ and $[v]$ as given in Definition 8.2.5, F is a group, called the **free group** on the set X.

PROOF   From the Definition 8.2.3, it is clear that for the *words* u, v, w, we have $u(vw) = (uv)w$. This means that $[u]([v][w]) = [u][vw] = [u(vw)] = [(uv)w] = [uv][w] = ([u][v])[w]$.

Thus the associative rule holds for the product in F.

The identity element of F is $[e]$ and the inverse of $[y_1 y_2 y_3 \ldots y_t]$ is

$$[y_t^{-1} y_{t-1}^{-1} y_{t-2}^{-1} y_{t-3}^{-1} \ldots y_1^{-1}],$$

where $y_i \in \overline{X}$ and $(x_i^{-1})^{-1}$ is taken to be x_i.

Thus F is a group according to Definition 8.2.1.

Just as we usually take the fraction $2/3$ to represent the equivalence class $[2/3] = \{2/3, 8/12, \ldots\}$ of all fractions equivalent to $2/3$, so it is usual to take $y_1 y_2 y_3 \ldots y_t$ as the representative of the equivalence class (set of equivalent *words*) $[y_1 y_2 y_3 \ldots y_t]$, where, in $y_1 y_2 y_3 \ldots y_t$, there is *no* occurrence of pairs like

$$x_i x_i^{-1} \quad \text{or} \quad x_i^{-1} x_i.$$

Thus, in Example 8.2.2, we would usually take $x_2 x_4^{-1} x_4^{-1}$ to represent the equivalence class containing all the *words* listed there.

A *word* such as $x_2 x_4^{-1} x_4^{-1}$ is said to be **reduced**. It corresponds to a fraction in its lowest terms in our comparison of equivalence classes of *words* with equivalence classes of fractions.

In order to extend the above theory to describe not just free groups but *all* groups, we let $w = y_1 y_2 y_3 \ldots y_t$ be *any word*, where $y_i \in \overline{X}$. We then say that two *words* u and v are **equivalent** if they differ only in the presence of the word w, or pairs of symbols like $x_i x_i^{-1}$ or $x_i^{-1} x_i$.

For example, if $w = x_1 x_2^{-1} x_3$, then $x_1 x_2^{-1} x_3 x_1^3 x_2 x_4 x_1 x_2^{-1} x_3 x_1 x_2 x_4 x_1 x_1^{-1}$ is equivalent to $x_1^3 x_2 x_4 x_1 x_2 x_4$.

Now we argue along lines exactly similar to those for the free group. The group so defined is denoted by

$$G = \langle x_1, x_2, \ldots, x_n | w = e \rangle$$

and is said to be **generated** by the **generators** $x_1, x_2, \ldots, x_n$, subject to the **relation** $w = e$. It is not stated explicitly that we also have the relations $x_i x_i^{-1} = e$ and $x_i^{-1} x_i = e$ for $i = 1, 2, 3, \ldots, n$. These relations are sometimes called the **trivial** relations.

We can extend the above to any finite number of relations $w_1 = e$, $w_2 = e$, $\ldots$, $w_r = e$ and obtain the group written as

$$G = \langle x_1, x_2, x_3, \ldots, x_n | w_1 = e, \ldots, w_r = e \rangle$$

Not all groups can be described in this way, but those groups that can are said to be **finitely presented**. However, if we allow the number of generators and the number of relations to be infinite, then every group can be described in terms of generators and relations. Groups of interest in topology usually occur naturally in this way, as we shall see a little later.

Let us consider an example.

EXAMPLE 8.2.3   Let

$$G = \langle x_1, x_2 | x_1^2 = e,\ x_2^2 = e,\ (x_1 x_2)^2 = e \rangle$$

Thus G is generated by two elements x_1 and x_2 subject to the relations $x_1^2 = e$, $x_2^2 = e$, and $(x_1 x_2)^2 = e$.

This means that every time x_1^2 occurs in a *word* we can replace it by e and still represent the *same* element of the group G. Similar observations apply to the other relations. Thus, effectively, G has just four distinct elements, namely

$$e, x_1, x_2, x_1 x_2.$$

To show how these elements are multiplied, it is convenient to give a multiplication (or Cayley) table as follows.

Table 8.2.1

	e	x_1	x_2	$x_1 x_2$
e	e	x_1	x_2	$x_1 x_2$
x_1	x_1	e	$x_1 x_2$	x_2
x_2	x_2	$x_1 x_2$	e	x_1
$x_1 x_2$	$x_1 x_2$	x_2	x_1	e

From this table we can easily see that the inverse of x_1 is x_1, and so on.

There are many interesting properties of multiplication tables of groups. For example, each element of the group appears once and only once in each row and in each column of the table. Many other facts of this kind are given in Grossman & Magnus [15], where a connection between groups and graphs is also discussed. This has some relevance to Chapter 6 of this book. Nevertheless we shall not pursue the connection here.

The branch of group theory which is concerned with groups described in terms of generators and relations is called **combinatorial group theory**. Apart from Johnson [18] already mentioned, standard texts on this topic are Lyndon & Schupp [22] and Magnus, Karrass & Solitar [24].

8.3 PLANE HYPERBOLIC GEOMETRY

Although there are several kinds of *non*-Euclidean geometry, the term is usually used to mean *hyperbolic geometry*. A readable account of how the subject arose is given in Meschkowski [26]. A somewhat more detailed account is given by Bonola [3]. Limitations of space restrict us to a brief survey.

Euclid's fifth axiom, usually called the *Parallel Postulate*, may be stated conveniently in a form due to Playfair (1795) as follows.

If l is any line and p is any point not on l, then we can construct a unique line through p which is parallel to l.

Note that parallel lines are defined to be lines that *do not meet*.

For many years mathematicians tried to prove that this axiom was a consequence of Euclid's four other basic postulates as listed in section 8.1. They failed. Eventually, through the efforts, in particular, of Gauss, Bolyai, and Lobachevsky, it came to be realized that a completely new geometry could be constructed satisfying all Euclid's postulates excepting the fifth, which is negated. This, incidentally, showed that the fifth postulate is independent of the others. In the new geometry in place of the fifth postulate we have:

If l is any line and p is any point not on l, then we can construct at least two distinct lines through p which are parallel to l.

Exactly as with Euclidean plane geometry, five axioms are not sufficient to characterize the geometry, and Hilbert again provided a complete set of axioms. This non-Euclidean geometry is often referred to as *hyperbolic geometry*.

In the Euclidean case in section 8.1, we took the usual coordinate geometry of the plane as a model of axiomatic Euclidean geometry. In this way we avoided working directly from the axioms. We follow the same approach for hyperbolic geometry.

The objects of hyperbolic geometry, just as for Euclidean geometry, are *points* and *lines*. However, it is important to realize that these objects, although called points and lines, may be *any objects that satisfy the axioms*. With this in mind we give the following Poincaré model of hyperbolic geometry; see Greenberg [14] and Sawyer [28]. Of course, it should be checked that this model does satisfy all the axioms of hyperbolic geometry. However, in this elementary treatment we ask the reader to assume that this can be done.

We represent hyperbolic geometry on the *interior* of a fixed disc of unit radius denoted by D. **Points** in hyperbolic geometry are just points in the disc, but *not* on the boundary circle C, or outside it. **Lines** in hyperbolic geometry are the arcs of circles inside the disc which meet the boundary circle at right-angles; in particular, the diameters of the disc are lines. The hyperbolic **angle** between two hyperbolic lines which meet at a point is the ordinary Euclidean angle between their tangents. In this section, points, lines, etc., will refer to hyperbolic points, lines, etc. If we mean point or line in the ordinary Euclidean sense, we shall write Euclidean point (abbreviated E-point), Euclidean line (abbreviated E-line), and so on.

Two points lie on a line if and only if they lie inside D on an E-circle cutting C at right-angles.

Hyperbolic angles are equal if and only if they are equal as Euclidean angles.

Fig. 8.3.1 illustrates these ideas.

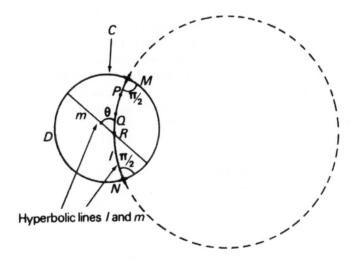

Fig. 8.3.1.

Points P, Q, and R lie on the hyperbolic line l. Hyperbolic lines l and m intersect at R. θ is the hyperbolic angle between lines l and m. The points M and N at which the E-arc l cuts the E-circle C are *not*, of course, *on* the line l. They are the so-called 'points at ∞' of the line l.

It is possible to define the distance between two points P and Q in hyperbolic geometry. Let $E(PN)$ denote the Euclidean distance between the points P and N along the *chord* joining P and N. Then the **hyperbolic distance** between the points P and Q shown in Fig. 8.3.1 is defined to be:

$$(1/k) \left| \log_e \frac{E(PN)E(QM)}{E(PM)E(QN)} \right|,$$

where k is twice the radius of C. Thus in our model, where C has unit radius, k is 2 and we have:

$$\text{distance } PQ = \frac{1}{2} \left| \log_e \frac{E(PN)E(QM)}{E(PM)E(QN)} \right|.$$

The distance PQ increases without limit as P and Q approach the circle C. This means that lines in hyperbolic geometry have infinite length, as we would expect since the first four postulates of Euclid are satisfied.

In section 8.1, we presented the point of view, due to Klein, that Euclidean plane geometry is the study of those properties of the plane that are unchanged under the isometries of the Euclidean group E_2. In a similar way, it can be shown that we can regard hyperbolic geometry as the study of those properties of the hyperbolic plane, represented in our model by the interior of the unit disc D, that are unchanged under those transformations of the hyperbolic plane which preserve the hyperbolic distance between two points, i.e. hyperbolic isometries. The latter form the elements of the hyperbolic group H_2. In the Euclidean case, E_2 is generated by reflections. In a similar way, the hyperbolic group is generated by *hyperbolic reflections*. In order to describe the special hyperbolic isometries that we call hyperbolic reflections, we have to define a certain kind of transformation of part of the Euclidean plane $\mathbb{R}^2$ called an *inversion in a circle*.

Let I denote the **circle of inversion**, centre O, radius r. Let f denote the **inversion** in the circle I defined as follows.

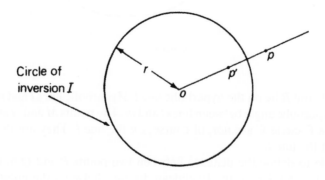

Fig. 8.3.2.

Take any point p of the Euclidean plane other than O. Join O to p. Then $f(p)$ is defined to be that point p' lying in the E-line Op such that $E(Op)E(Op') = r^2$.

Of course, since f is not defined on O, f is not a transformation of the whole plane $\mathbb{R}^2$. But this does not matter for our purpose.

In Fig. 8.3.3 we show a typical hyperbolic line l and point p.

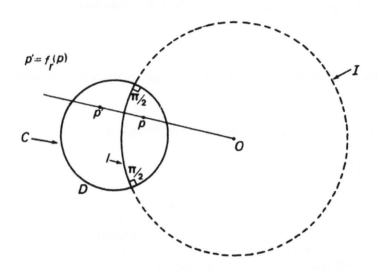

Fig. 8.3.3 — The construction of a hyperbolic reflection f_r in the line l.

We define a **hyperbolic reflection** f_r in the hyperbolic line l to be the restriction to the disc D of an inversion f in the circle I of which the line l is a part.

Note that any point of D is transformed into another point of D as required.

This transformation f_r of D behaves as we would expect a reflection to behave, in the sense that all points on one side of the line l are transformed into points on the other side of l, and all points on the line l are left fixed. In fact, although we do not show it here, using the definition of distance given earlier, it can be proved that the points p and p' are 'mirror' images of each other in the hyperbolic line l.

Just as in Euclidean plane geometry, we can use the isometries of the hyperbolic plane to prove theorems in hyperbolic geometry. For example, assuming that we have already proved:

(1) given any two points p and q in D, there is an isometry $f \in \mathbf{H}_2$ such that $f(p) = q$,
(2) every isometry preserves angles between intersecting lines,
(3) every isometry takes lines into lines,

then we can prove

THEOREM 8.3.1   Every (hyperbolic) triangle has an angle sum less than π.

PROOF As shown in Fig. 8.3.4., let the vertices of the triangle be p_1, p_2, p_3 at which the angles are $\alpha_1, \alpha_2, \alpha_3$, respectively.

Choose $f \in H_2$ to send p_1 to O, the centre of D. Then the lines $p_1 p_2$ and $p_1 p_3$ lie along *diameters* of D, as shown. The angles of the triangle with vertices $O, f(p_2)$, $f(p_3)$ are *still* α_1, α_2, and α_3, because f preserves angles between intersecting lines.

But now, since α_2 and α_3 are certainly *less* than the corresponding angles of the *Euclidean* triangle with vertices $O, f(p_2), f(p_3)$, it follows that

$$\alpha_1 + \alpha_2 + \alpha_3 < \pi$$

Another interesting theorem in hyperbolic geometry, which we quote without proof, is

THEOREM 8.3.2   The area of an hyperbolic triangle with angles α, β, γ is

$$K(\pi - (\alpha + \beta + \gamma)),$$

where K is the *same* constant for *all* triangles.

By taking a suitable unit of area, K may be taken to be 1. Then the area of *any* triangle is $\pi - (\alpha + \beta + \gamma)$, which is often called the **angular defect** of the triangle.

For further details about hyperbolic geometry, the interested reader is referred to Greenberg [14] and to the references given there.

8.4   PLANE TESSELLATIONS

In this section we are concerned with covering the whole plane (Euclidean or hyperbolic) with a mosaic of polygons without overlapping. Such an arrangement is called a **tessellation** of the plane, or a **plane tessellation**. We shall be concerned with tessellations on both the Euclidean plane $\mathbb{R}^2$ and the hyperbolic plane as represented by the interior of the unit disc given in section 8.3. Of course, in the latter case, the polygons are hyperbolic polygons, i.e. their sides are hyperbolic lines. In particular, we shall consider *regular* tessellations. Formally we give.

DEFINITION 8.4.1   A **regular tessellation** of the Euclidean (hyperbolic) plane is a covering of the entire plane by *non*-overlapping Euclidean (hyperbolic) regular polygons meeting only along complete edges, or at vertices. All polygons in any one tessellation must have the same number of sides (edges). A regular tessellation with q regular p-sided polygons (p-gons) meeting at each vertex is denoted by $\{p, q\}$.

It follows immediately that the interior angle of a p-gon at a vertex must be $2\pi/q$.

In Fig. 8.4.1, we show a part of a regular polygon with p sides (a p-gon). AB is a side of the p-gon and O is its centre.

OA bisects the interior angle between two adjacent edges of the polygon as does OB.

Now $A\hat{O}B = 2\pi/p$, hence, for the Euclidean plane,

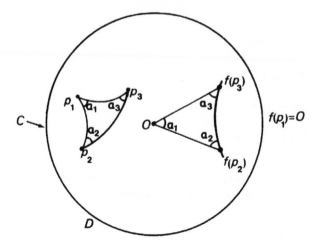

Fig. 8.3.4.

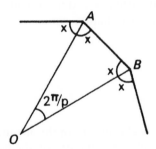

Fig. 8.4.1 — Part of a regular p-gon.

$$\pi - 2\pi/\mathrm{p} = O\hat{A}B + O\hat{B}A = 2O\hat{A}B$$

$\qquad$ = the angle between two adjacent edges of the polygon.

$\quad$ For a regular tessellation, the angle between adjacent edges of a p-gon equals the interior angle of a p-gon at a vertex, as shown in Fig. 8.4.2.

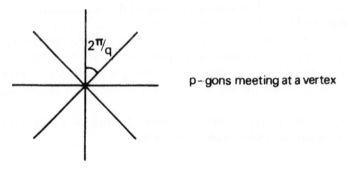

Fig. 8.4.2.

Thus

$$\pi - 2\pi/p = 2\pi/q\,.$$

Hence

$$1 - 2/p = 2/q\,.$$

Thus

$$pq - 2q - 2p = 0\,.$$

Hence

$$(p-2)(q-2) = 4\,.$$

There are *three real* solutions of this equation. Hence there are just *three* regular tessellations on the Euclidean plane $\mathbb{R}^2$, namely

$$\{4,4\},\ \{6,3\}\ \text{and}\ \{3,6\}\,.$$

On the hyperbolic plane, by Theorem 8.3.1, the sum of the angles of a triangle is less than π. Hence the angle between two adjacent edges of the hyperbolic p-gon is less than $\pi - 2\pi/p$. For a regular tessellation, similar arguments to those above give.

$$2\pi/q < \pi - 2\pi/p\,.$$

Hence

$$(p-2)(q-2) > 4\,.$$

There is an *infinite* number of real solutions of this inequality. Therefore there is an *infinite* number of possible regular tessellations on the hyperbolic plane.

Note, in particular, that if $p = q$, then we may take p to be any positive integer greater than 4.

With a tessellation on a plane, we can associate a group of isometries (Euclidean or hyperbolic), namely the group of all **symmetries** of the tessellation. This is the group of all those isometries of the plane which leave the tessellation *apparently* unchanged.

For example, if we have the tessellation $\{p,q\}$, made up of p-gons, q at each vertex, then reflections r_1, r_2, r_3 in the three sides of the triangle ABC shown in Fig. 8.4.3. generate the complete symmetry group of the tessellation.

Note that the lines of reflection *stay fixed* in space throughout. Also note that reflection in a side means reflection in the line of which the side is part.

Suppose that g is an element of a group, and that $g^n = e$ (the identity of the group), but $g^r \neq e$, if $0 < r < n$. Then we say that g is an element of **order** n, and we write $o(g) = n$. If $g^n \neq e$ for any n, then we say that g has **infinite order**, written $o(g) = \infty$.

Reflections are clearly elements of order 2. Such elements are often called **involutions**. In our particular example, we have

$$r_1^2 = r_2^2 = r_3^2 = e\,.$$

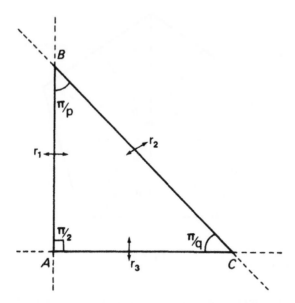

Fig. 8.4.3.

Now $r_2 r_1$ (meaning first r_1 then r_2) rotates the triangle ABC through an angle $2\pi/p$ in an anti-clockwise sense. (Remember that r_2 is a reflection of the triangle, in the position it is in as a result of r_1, in the side BC of the 'original' triangle ABC.) Hence $(r_2 r_1)^p$ rotates the triangle through 2π. It follows that $(r_2 r_1)^p = e$ and $o(r_2 r_1) = p$. Similarly $(r_3 r_2)^q = e$ and $(r_1 r_3)^2 = e$.

Note that it is a useful fact that if in a group $(ab)^n = e$, then $(ba)^n = e$.

Although we do not prove it here, these relations suffice to determine the complete symmetry group of the tessellation $\{p, q\}$. Thus, according to section 8.2, this group can be described as

$$\langle r_1, r_2, r_3 | r_1^2 = r_2^2 = r_3^2 = (r_2 r_1)^p = (r_3 r_2)^q = (r_1 r_3)^2 = e \rangle$$

Let us denote this group by $[p, q]$. (See [6].)

The relation between the triangle ABC and the associated p-gon is shown in Fig. 8.4.4 for the case when $p = 5$.

Starting with the triangle ABC and transforming it using just the reflections r_1 and r_2, we obtain the regular p-gon which is the so-called **fundamental region** of the regular tessellation $\{p, q\}$. The group

$$\langle r_1, r_2 | r_1^2 = r_2^2 = (r_2 r_1)^p = e \rangle$$

is the **symmetry group** of the regular p-gon. It is usually called the **dihedral group** D_p. The number of elements in D_p, called the **order** of D_p, is 2p.

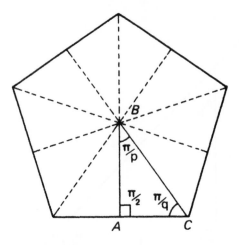

Fig. 8.4.4.

Note that D_p is a **subgroup** of $[p,q]$ i.e. a group contained in $[p,q]$ with the *same product* as in $[p,q]$.

We *could* take triangle ABC as our fundamental region and transform it using the isometries of $[p,q]$ to cover the whole plane with triangles like ABC. This would give a tessellation of the plane but *not* a regular one.

EXAMPLE 8.4.1   We consider the tessellation $\{p,q\}$ of the Euclidean plane $\mathbb{R}^2$ when $p = q = 4$. This is a covering of the plane with squares. The symmetry group $[4,4]$ is described by

$$\langle r_1, r_2, r_3 | r_1^2 = r_2^2 = r_3^2 = (r_2 r_1)^4 = (r_3 r_2)^4 = (r_1 r_3)^2 = e \rangle,$$

where r_1, r_2, r_3 are reflections in the sides (extended) of the triangle ABC as shown in Fig. 8.4.5. Recall that the lines of reflection *stay fixed* in space.

From the reflections r_1 and r_2 we generate the fundamental region, the square, shown in Fig. 8.4.6.

In each triangular region $\triangle_i, i = 1, 2, 3, 4, 5, 6, 7, 8$, we have written the isometry which transforms the initial triangle ABC into the triangle $\triangle_i$.

Note that any particular isometry can be written in many different ways in terms of r_1 and r_2. That they represent the *same* isometry can be shown by using the relations in the group D_4 generated by r_1 and r_2. This dihedral group is the symmetry group of a square. We have

$$D_4 = \langle r_1, r_2 | r_1^2 = r_2^2 = (r_2 r_1)^4 = e \rangle$$

The number of elements in D_4, i.e. the number of *distinct* symmetries of the square, is 8. We write $|D_4| = 8$.

We can cover the plane with squares by using two translations a_1 and a_2 and their inverses. The translations shift the fundamental region $CDEF$ along directions

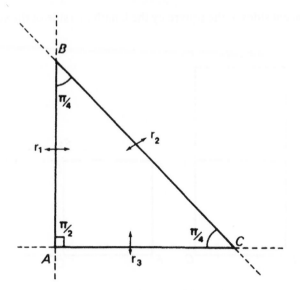

Fig. 8.4.5.

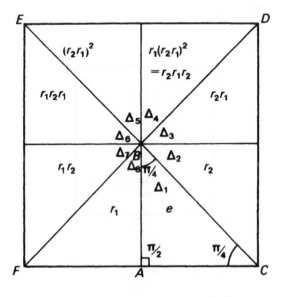

Fig. 8.4.6 — Fundamental region for {4,4}.

parallel to adjacent sides of the square by the length of a side of the square, as shown in Fig. 8.4.7.

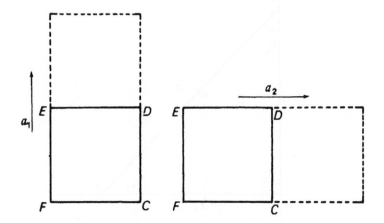

Fig. 8.4.7 — Translations a_1 and a_2.

We can write a_1 and a_2 in terms of the reflections r_1, r_2, and r_3 as follows.

$$a_1 = r_2 r_1 r_2 r_3 \quad \text{and} \quad a_2 = r_2 r_1 r_3 r_2 (r_1 r_2)^2.$$

The reader is invited to check that these isometries a_1 and a_2 do indeed transform the fundamental region $CDEF$ as stated. This can conveniently be done by marking out the square in Fig. 8.4.8 on a cut-out model, and performing the appropriate transformations on this model.

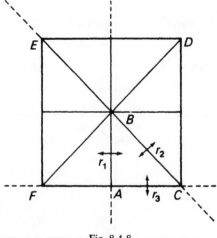

Fig. 8.4.8.

In carrying out these transformations, remember that the lines of reflection stay fixed in space.

In order to obtain the dual tessellation and relate it to a plane model of the torus T, it is convenient to have available the idea of a Cayley graph of a group.

Suppose we have a group G generated by the elements $x_1, x_2, x_3, \ldots$. For each distinct element g_i of the group, we have a vertex of the Cayley graph D. Vertex g_i is connected to vertex g_k by a *directed* edge from g_i to g_k if and only if $g_k = x_j g_i$ for *some generator* x_j. The edge is labelled with some symbol denoting the generator x_j. This symbol could be x_j itself. Fig. 8.4.9 shows the Cayley graph for the Klein four-group

$$\langle x_1, x_2 | x_1^2 = x_2^2 = (x_1 x_2)^2 = e \rangle$$

considered in Example 8.2.3. It should be noted that the Cayley graph for a given group G depends on the *generators* used to obtain the group. Thus the same group has many different representations as a Cayley graph.

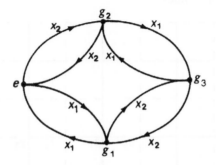

Fig. 8.4.9 — A Cayley graph for the Klein four-group.

For the elements of the Klein four group, we write e, $g_1 = x_1$, $g_2 = x_2$ and $g_3 = x_1 x_2$.

Traversing an edge x_j in the 'wrong' direction is associatd with the action of x_j^{-1}. Often with involutory generators, like x_1 and x_2 above, each pair of directed edges corresponding to the involutory generator is replaced by a single undirected edge, as shown in Fig. 8.4.10 for the Klein four-group.

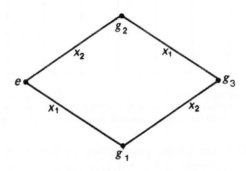

Fig. 8.4.10.

Let us now apply these ideas to the tessellation $\{4,4\}$ obtained by applying the two translations a_1 and a_2 to the fundamental region $CDEF$. Of course to obtain the complete tessellation we use the translations and their inverses over and over again. In fact, we use all the isometries in the group $\pi(T)$ generated by a_1 and a_2. The latter is a subgroup of $[4,4]$, which incidentally shows that $[4,4]$ is *infinite*, because translations have infinite order. On the original tessellation $\{4,4\}$ of the plane obtained by translations of the fundamental region $CDEF$, we superimpose the Cayley graph D_T for $\pi(T)$. Each vertex v_i of D_T is placed in the centre of that square of the tessellation produced by applying the isometry v_i to the fundamental region $CDEF$. Then a directed straight line edge is drawn from the vertex v to the vertex w if $w = a_1v$ or if $w = a_2v$. This results in a Cayley graph describing the dual tessellation, in this case another example of the original tessellation $\{4,4\}$. Fig. 8.4.11 shows a part of the original tessellation together with the dual tessellation. The former is shown drawn with broken lines.

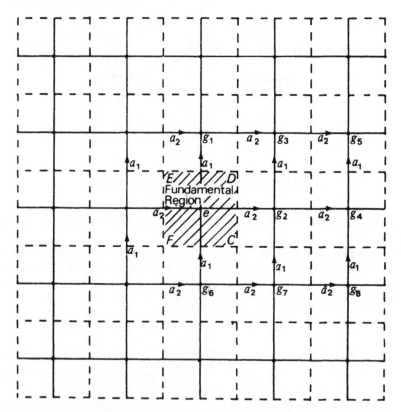

Fig. 8.4.11 — Here we have $g_1 = a_1, g_2 = a_2, g_3 = a_2a_1, g_4 = a_2^2, g_5 = a_2^2a_1, g_6 = a_1^{-1}, g_7 = a_2a_1^{-1}$, and $g_8 = a_2^2a_1^{-1}$. The g_i label the vertices of the Cayley graph. The original tessellation is shown in broken lines. Both the original tessellation and the dual are examples of $\{4,4\}$.

Any closed path in the Cayley graph gives a relation which is satisfied in the associated group. For example, in Fig. 8.4.11, if we start at the vertex e and follow a path through the vertices e, a_1, a_2a_1, a_2, and back to e, we see that the relation

$$a_2^{-1} a_1^{-1} a_2 a_1 = e$$

is satisfied in the group $\pi(T) = \langle a_1, a_2 \rangle$. The expression $a_2^{-1} a_1^{-1} a_2 a_1$ (notice this is the *word* representation of T) is often written $[a_2, a_1]$ and is called the **commutator** of a_2 and a_1. The relation $[a_2, a_1] = e$ implies that a_2, and a_1 **commute**, i.e. $a_2 a_1 = a_1 a_2$.

Note that when we move along an edge of the Cayley graph in a direction opposite to that associated with the edge, we use the inverse of the appropriate generator.

Since $\pi(T)$ is a *subgroup* of $[4,4]$, we would expect the relation $[a_2, a_1] = e$ to be a consequence of the relations defining $[4,4]$. That this is indeed the case is shown below.

In $[4,4]$, we have $r_1^2 = r_2^2 = r_3^2 = e$, and $(r_2 r_1)^4 = (r_3 r_2)^4 = (r_1 r_3)^2 = e$. We also note the following facts.

If, in a group, $(xy)^n = e$ then $(yx)^n = e$.
From $(xy)^2 = x^2 = y^2 = e$, we deduce that $xy = yx$.
For the latter the manipulations are:

$$xyxy = e \rightarrow xyxyy = y \rightarrow xyx = y \rightarrow xyxx = yx \rightarrow xy = yx.$$

For the former we have $y(xy)^n x = yex$, and so $(yx)^n yx = yx$.
Thus $(yx)^n yxx^{-1} y^{-1} = yxx^{-1} y^{-1}$. Hence $(yx)^n = e$.

Finally $(xy)^{-1} = y^{-1} x^{-1}$.
This follows by direct verification.

$$
\begin{aligned}
a_2^{-1} a_1^{-1} a_2 a_1 &= (r_2 r_1)^2 r_2 r_3 r_1 r_2 \cdot r_3 r_2 r_1 r_2 \cdot r_2 r_1 r_3 r_2 (r_1 r_2)^2 \cdot r_2 r_1 r_2 r_3 \\
&= (r_2 r_1)^2 r_2 r_3 r_1 r_2 (r_3 r_2)^2 r_1 r_3 \\
&= (r_1 r_2)^2 r_2 r_1 r_3 r_2 (r_2 r_3)^2 r_3 r_1 \quad * \\
&= r_1 r_2 r_3 r_2 r_2 r_3 r_2 r_1 \\
&= e.
\end{aligned}
$$

In * we have used the fact that $(r_2 r_1)^2 = (r_1 r_2)^2$, and $(r_3 r_2)^2 = (r_2 r_3)^2$. The former follows from $(r_2 r_1)^4 = e$. We have $(r_2 r_1)^2 r_2 r_1 r_2 r_1 = e$, and hence

$$
\begin{aligned}
(r_2 r_1)^2 &= (r_2 r_1 r_2 r_1)^{-1} = r_1^{-1} r_2^{-1} r_1^{-1} r_2^{-1} \\
&= r_1 r_2 r_1 r_2 = (r_1 r_2)^2.
\end{aligned}
$$

The other relation follows in an exactly similar way.

Although we do not prove it here, the relation $[a_2, a_1] = e$ is sufficient to define $\pi(T)$. This means that

$$\pi(T) = \langle a_1, a_2 \mid a_2^{-1} a_1^{-1} a_2 a_1 = e \rangle,$$

and all other relations satisfied in $\pi(T)$ can be deduced (in theory) from $[a_2, a_1] = e$, or the equivalent relation $a_2 a_1 = a_1 a_2$, and, of course, the trivial relations

$$a_1 a_1^{-1} = a_1^{-1} a_1 = e, \quad \text{and} \quad a_2^{-1} a_2 = a_2 a_2^{-1} = e,$$

which hold in every group.

Thus $\pi(T)$ is an infinite abelian group. In fact, it is a free abelian group on two free generators. As such it is a direct product of two infinite cyclic groups i.e. groups isomorphic to the integers under ordinary addition (see Gardiner [10] for further details). $\pi(T) = \{a_1^n a_2^m | n, m \text{ any integers}\}$.

Now consider any face of the dual tessellation or Cayley graph. The edges labelled with the generators a_1 and a_2 give the diagram shown in Fig. 8.4.12.

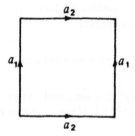

Fig. 8.4.12.

It was following a path round such a face that gave us the relation $a_2^{-1} a_1^{-1} a_2 a_1 = e$ defining $\pi(T)$.

Dropping the arrows and indicating direction by using inverses, we obtain Fig. 8.4.13.

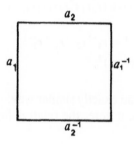

Fig. 8.4.13.

This looks very familiar. It can be interpreted as a plane model of a torus T. The space model obtained from this plane model is shown in Fig. 8.4.14, together with the loops formed by the edges a_1 and a_2.

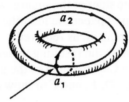

Point p at which loops
intersect

Fig. 8.4.14.

The group generated by a_1 and a_2, with product defined as one loop followed by another, always starting from some fixed point p, say, at which the loops may be taken to intersect, is called the **fundamental group** of the torus. Such a group can be defined for any surface and is of central importance in algebraic topology. We shall be considering it again in Chapter 10. In the case of the torus, it turns out to be precisely the group.

$$\pi(T) = \langle a_1, a_2 | a_2^{-1} a_1^{-1} a_2 a_1 = e \rangle,$$

which generates the tessellation $\{4,4\}$ from the fundamental region $CDEF$.

We call this tessellation and the associated group $\pi(T)$, a **tessellation representation** of the torus T.

EXAMPLE 8.4.2   If we obtain the same tessellation $\{4,4\}$ by using different transforming isometries, we can obtain a tessellation representation of the Klein bottle K called the universal covering space of K. We outline below how this can be done and leave the details to the reader.

The isometries a_1 and a_2 used in Example 8.4.1 each contain an even number of reflections and leave each square in the tessellation with the same face uppermost. By using an isometry with an *odd* number of reflections, we could obtain a tessellation in which some squares are 'flipped over', e.g. a **glide reflection**, where we follow a translation with a reflection in the line of the translation, or conversely. With the same notation as in Figs. 8.4.5 and 8.4.6, suitable isometries are

$$b_1 = r_1 a_1 = r_1 r_2 r_1 r_2 r_3 = (r_1 r_2)^2 r_3$$

and

$$b_2 = a_2 = r_2 r_1 r_3 r_2 (r_1 r_2)^2 = r_2 r_3 r_2 r_1.$$

The latter simpler expression for a_2 can be obtained by using the relations $(r_1 r_2)^2 = (r_2 r_1)^2$ and $r_1 r_3 = r_3 r_1$, obtained in Example 8.4.1.

The isometry b_1 moves the fundamental region $CDEF$ up by the length of its side and then reflects in the line AB of the triangle ABC. This essentially involves 'identifying' two opposite sides of the square, with a 'twist', in the same way that two opposite edges of a strip of paper are identified to make a Möbius strip. A touch of *non*-orientability is creeping in, as required!

In Fig. 8.4.15, we show the resulting tessellation (in broken lines) and the dual tessellation, which is the Cayley graph of the group $\langle b_1, b_2 \rangle$. We have labelled each square with the letters C, D, E, and F to show how the fundamental region has been transformed.

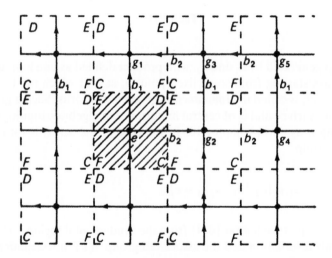

Fig. 8.4.15 — We denote the vertices of the Cayley graph or dual tessellation by g_i, where in the figure we show $g_1 = b_1$, $g_2 = b_2$, $g_3 = b_1 b_2$, $g_4 = b_2^2$, and $g_5 = b_1 b_2^2$. The broken lines give the original tessellation.

The isometry b_1 takes a square out 'through' the side ED and then 'identifies' that side with the side CF of the 'new' square. Similarly b_2 takes a square out through the side DC and identifies the latter with the side EF of the 'new' square.

Following the path round the periphery of a face of the dual tessellation, or Cayley graph, we obtain the relation $b_2^{-1} b_1^{-1} b_2^{-1} b_1 = e$, or equivalently, by taking the inverse of both sides of this relation, or by following the path in an anti-clockwise sense, $b_1^{-1} b_2 b_1 b_2 = e$.

As in Example 8.4.1, this can be obtained as a consequence of the relations in [4, 4].

As before, this relation suffices to define the group $\langle b_1, b_2 \rangle$, which is another subgroup of [4, 4].

The plane model associated with this relation is that of a Klein bottle, as shown in Fig. 8.4.16.

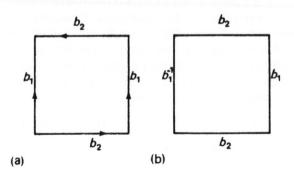

Fig. 8.4.16.

In Fig. 8.4.17, we show the corresponding space model of K with the edges b_1 and b_2 forming loops on K, crossing each other at the point p. Thus the fundamental group of K is

$$\pi(K) = \langle b_1, b_2 | b_1^{-1} b_2 b_1 b_2 = e \rangle .$$

As noted above, $\pi(K)$ is a subgroup of $[4,4]$.

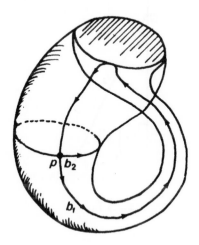

Fig. 8.4.17.

EXAMPLE 8.4.3   As we have seen in Examples 8.4.1 and 8.4.2, by selecting suitable isometries to transform the fundamental region, we can associate the tessellation $\{4,4\}$ with either a torus or a Klein bottle. Now the plane model of a torus, or a Klein bottle, is a 4-gon. Hence, if we want to obtain a tessellation representation of, say, a double torus, we expect our fundamental region to be an octagon. Moreover, to carry through the arguments of Examples 8.4.1 and 8.4.2, we want our tessellation to be self-dual. This means that for a double torus, we want a tessellation $\{8,8\}$. However, such a tessellation is not possible on a Euclidean plane according to our calculations at the beginning of this section. For this reason, we turn to the hyperbolic plane, where such tessellations are possible.

The complete symmetry group of the tessellation will be

$$[8,8] = \langle r_1, r_2, r_3 | r_1^2 = r_2^2 = r_3^2 = (r_1 r_2)^8 = (r_2 r_3)^8 = (r_3 r_1)^2 = e \rangle$$

and the fundamental region will be an octagon with symmetry group

$$D_8 = \langle r_1, r_2 | r_1^2 = r_2^2 = (r_1 r_2)^8 = e \rangle,$$

where r_1, r_2, and r_3 are *hyperbolic* reflections in the sides of a right-angled isosceles hyperbolic triangle OLE with its other angles $\pi/8$. The hyperbolic isometries r_1 and r_2 transform the triangle OLE into the fundamental region, the hyperbolic octagon, $ABCDEFGH$. As ususal, we represent the hyerbolic plane on the interior of the unit disc in accordance with section 8.3.

Figs 8.4.18 and 8.4.19 are the analogues of Figs. 8.4.5 and 8.4.6, with a similar notation.

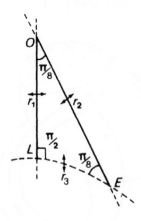

Fig. 8.4.18.

As before, the *hyperbolic* lines of reflection *stay fixed* in the hyperbolic plane.

Since the dual tessellation is to be of the same *form* as the original tessellation and each face is to give the plane model of a double torus, it follows that in our original tessellation the octagons should fit together so that pairs of sides are identified as in the plane model of a double torus, i.e. the edge, say EF, of one octagon is identified with the edge HG of its immediate neighbour. This would be in accord with the plane

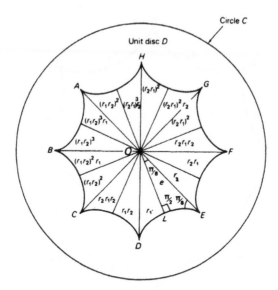

Fig. 8.4.19 — Fundamental region for the tessellation $\{8,8\}$.

model of the double torus shown in Fig. 8.4.20, where we have labelled the vertices as in the fundamental region.

Thus we have to look for hyperbolic isometries, expressed in terms of r_1, r_2, and r_3, which will send the fundamental region shown in Fig. 8.4.19 into octagons fitting together according to the above requirements.

It can be checked that the following isometries are suitable.

$$a_1 = r_3 r_1 (r_2 r_1)^2$$
$$a_2 = (r_1 r_2)^3 r_3 r_1 r_2 r_1$$
$$a_3 = (r_1 r_2)^4 r_3 r_2 r_1 r_2$$
$$a_4 = r_2 r_3 (r_1 r_2)^3$$

In Fig. 8.4.21, we show the tessellation $\{8,8\}$ produced by the above isometries acting on the fundamental region. (Recall that the lines of reflection stay fixed and $r_1 r_2$ means first apply r_2 and then apply r_1.)

We have also labelled a few of the faces of the tessellation with the isometry which produces that particular octagon from the fundamental region.

In Fig. 8.4.22, we have drawn the tessellation on the unit disc in such a way that one of the vertices of the tessellation occurs at the centre of the disc. The dual tessellation, or Cayley graph of $\langle a_1, a_2, a_3, a_4 \rangle$, now looks exactly like the original tessellation.

In terms of the generators a_1, a_2, a_3, a_4, the g_i are as follows.

$$g_1 = a_1, \qquad g_2 = a_4, \qquad g_3 = a_2 a_1,$$

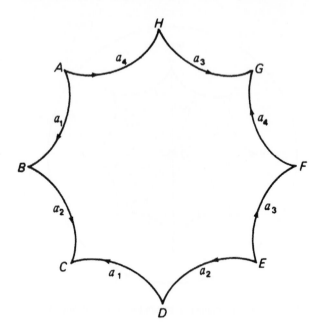

Fig. 8.4.20.

$$g_4 = a_3a_4, \qquad g_5 = a_1^{-1}a_2a_1,$$
$$g_6 = a_4^{-1}a_3a_4, \quad g_7 = a_2^{-1}a_1^{-1}a_2a_1.$$

Taking the isometries round the edges of a face of the dual tessellation, or Cayley graph, starting at vertex e, we obtain the relation

$$a_4^{-1}a_3^{-1}a_4a_3a_2^{-1}a_1^{-1}a_2a_1 = e$$

This gives us the plane model of a double torus as shown in Fig. 8.4.23. The corresponding space model is shown in Fig. 8.4.24, where the edges produce four loops intersecting at the point q. Thus the fundamental group of the double torus is given by

$$\pi(2T) = \langle a_1, a_2, a_3, a_4 | a_4^{-1}a_3^{-1}a_4a_3a_2^{-1}a_1^{-1}a_2a_1 = e \rangle,$$

which is a subgroup of $[8,8]$.

The relation $a_4^{-1}a_3^{-1}a_4a_3a_2^{-1}a_1^{-1}a_2a_1 = e$ is a consequence of the relations defining $[8,8]$. This can be checked as in Example 8.4.1.

The above theory can be generalized to a tessellation $\{4p, 4p\}$ on the hyperbolic plane representing pT. The complete symmetry group for the tessellation is

$$[4p, 4p] = \langle r_1, r_2, r_3 | r_1^2 = r_2^2 = r_3^2 = (r_1r_2)^{4p} = (r_2r_3)^{4p} = (r_3r_1)^2 = e \rangle.$$

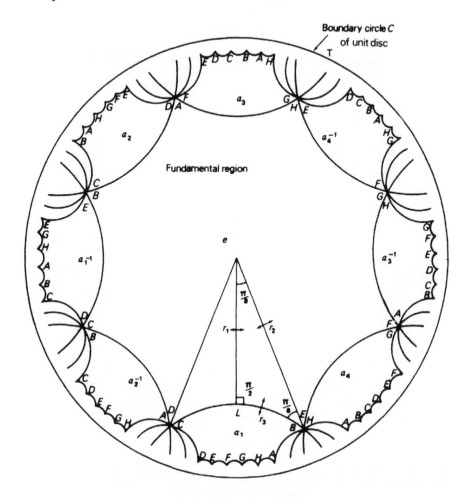

Fig. 8.4.21 — Tessellation {8,8} the universal covering space of the hyperbolic plane represented on the interior of a unit disc. The faces of the tessellation are labelled with the same letters as the fundamental region to indicate how the latter has been transformed.

The fundamental region is a 4p-gon with symmetry group

$$D_{4p} = \langle r_1, r_2 | r_1^2 = r_2^2 = (r_1 r_2)^{4p} = e \rangle,$$

where r_1, r_2, r_3 are hyperbolic reflections in the sides of a hyperbolic triangle with angles $\pi/2$, $\pi/4p$, and $\pi/4p$, as shown in Fig. 8.4.25.

The isometries producing the tessellation {4p, 4p} are

$$a_{2i-1} = (r_1 r_2)^{4i-4} r_3 r_1 (r_2 r_1)^{4i-2}$$

and

$$a_{2i} = (r_1 r_2)^{4i-1} r_3 r_1 (r_2 r_1)^{4i-3}, \quad i = 1, 2, \ldots, p.$$

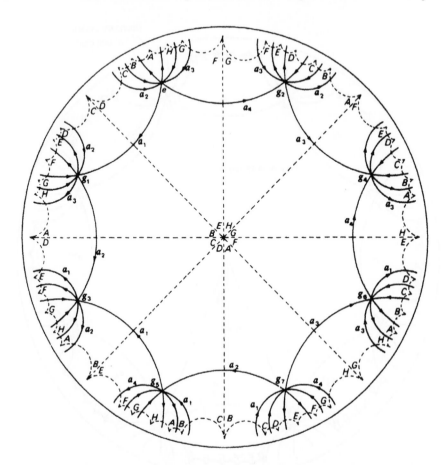

Fig. 8.4.22. — Tessellation $\{8,8\}$ shown in broken lines on the hyperbolic plane. The vertices of the dual tessellation, or Cayley graph, are labelled g_i. Expressions for the g_i in terms of the generators are given in the text.

They satisfy the relation

$$a_1 a_2 a_1^{-1} a_2^{-1} \ldots a_{2p-1} a_{2p} a_{2p-1}^{-1} a_{2p}^{-1} = e,$$

which is associated with the plane model of the connected sum of p tori.

The corresponding loops produced on the space model of pT generate the fundamental group $\pi(\mathrm{p}T)$ of pT. Thus

$$\pi(\mathrm{p}T) = \langle a_1, a_2, \ldots, a_{2p} \,|\, a_1 a_2 a_1^{-1} a_2^{-1} \ldots a_{2p-1}^{-1} a_{2p}^{-1} = e \rangle,$$

which is a subgroup of [4p, 4p].

The relation $a_1 a_2 a_1^{-1} a_2^{-1} \ldots a_{2p-1} a_{2p} a_{2p-1}^{-1} a_{2p}^{-1} = e$ is a consequence of the relations in the group [4p, 4p].

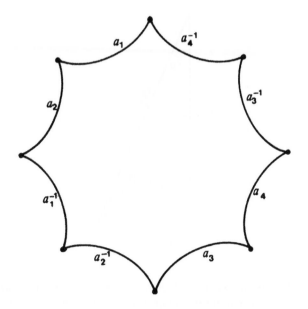

Fig. 8.4.23.

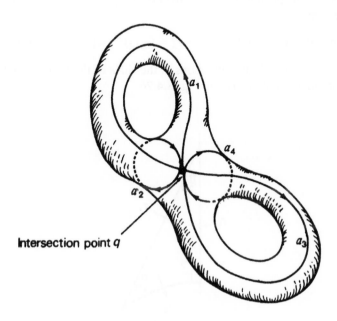

Intersection point *q*

Fig. 8.4.24.

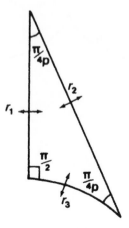

Fig. 8.4.25.

EXAMPLE 8.4.4 Just as we introduced *non*-orientability in Example 8.4.2 by using a glide reflection which 'flipped over' some of the polygons of the tessellation, so we can use the same idea to introduce *non*-orientability into the arguments in Example 8.4.3. In this way, we can find tessellation representations on the hyperbolic plane of the connected sum of n projective planes, *provided* n > 2. We outline the results below, and invite the reader to check them for, say, n = 3, and to draw the associated tessellations, both original and dual.

The tessellation produced on the hyperbolic plane is {2n, 2n} with complete symmetry group

$$[2n, 2n] = \langle r_1, r_2, r_3 \,|\, r_1^2 = r_2^2 = r_3^2 = (r_1 r_2)^{2n} = (r_2 r_3)^{2n} = (r_3 r_1)^2 = e \rangle \,,$$

where r_1, r_2, r_3 are hyperbolic reflections in the sides of the hyperbolic triangle with angles $\pi/2$, $\pi/2n$, and $\pi/2n$ shown in Fig. 8.4.26.

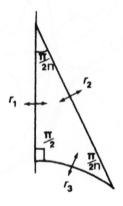

Fig. 8.4.26.

The fundamental region is a 2n-gon, with symmetry group

$$D_{2n} = \langle r_1, r_2 | r_1^2 = r_2^2 = (r_1 r_2)^{2n} = e \rangle,$$

produced by applying the isometries r_1 and r_2 to the triangle in Fig. 8.4.26.

The isometries which produce the tessellation $\{2n, 2n\}$ are

$$a_i = (r_1 r_2)^{2i-2} r_3 (r_2 r_1)^{2i-1}, \quad i = 1, 2, \ldots, n.$$

They satisfy the relation

$$a_1^2 a_2^2 a_3^2 a_4^2 \ldots a_n^2 = e,$$

which is associated with the plane model of nP.

The corresponding loops produced on the space model by the edges a_i generate the fundamental group $\pi(nP)$ of nP. Thus

$$\pi(nP) = \langle a_1, a_2, \ldots, a_n | a_1^2 a_2^2 \ldots a_n^2 = e \rangle,$$

for $n > 2$.

This is a subgroup of $[2n, 2n]$, and the relation $a_1^2 a_2^2 a_3^2 \ldots a_n^2 = e$ is a consequence of the relations in the group $[2n, 2n]$.

From the Examples 8.4.1–8.4.4 and the Classification Theorem 3.6.1, we deduce that we can find tessellations on either the Euclidean plane or the hyperbolic plane to represent *any* orientable compact surface *except* the sphere. Similarly, bearing in mind that K and $2P$ are homeomorphic, we can find tessellations on the Euclidean or hyperbolic planes to represent *any* *non*-orientable compact surface *except* the projective plane P.

Tessellations to represent the sphere and projective plane can be found on the *elliptic plane*. This involves a plane geometry different from both Euclidean and hyperbolic. Klein gave a model of elliptic geometry in which the lines are great circles on a sphere and a point is a *pair* of diametrically opposite 'ordinary' points on the sphere. In this geometry there are *no* parallel lines! (See the exercises for an alternative *disc* model.)

The reader is invited to investigate the possibility of reversing the above procedure by starting with a plane model of a compact surface, and then fitting the plane models together so as to obtain a tessellation of the plane, Euclidean or hyperbolic as the case may be. This produces our dual tessellation as the original tessellation.

8.5 COMMENTS

The theme which runs through this chapter and which unites its various parts is that of a group of transformations. Cayley in 1859 and, later Klein, related metric geometry, i.e. geometry involving the concept of distance between any two points, to projective geometry, which we mentioned in section 2.6. This work involved characterizing the various geometries according to their metric and non-metric properties. These ideas led Klein to consider the wider possibility of characterizing geometries by what they essentially were trying to accomplish. He introduced these ideas in a lecture in 1872, entitled *Vergleichende Betrachtungen uber neuere geome-trische Forschungen*, given on the occasion of his admission to the faculty of the University of Erlangen. For this reason, the views he expressed are often known as Klein's *Erlanger Programm*.

The essential idea is that each geometry can be characterized by a group of transformations, and is just the study of those properties of 'space' which are unchanged by transformations belonging to the group concerned. Thus topology is the study of those properties of space which are left fixed by continuous transformations. Projective geometry is the study of those properties of space unchanged by the so-called 'projective' transformations, and so on. In this way the group *defines* the geometry.

In recent years Coxeter and others have used the geometry to obtain information about the associated group of transformations. In this way many otherwise intractable problems, particularly about groups defined in terms of generators and relations, have been solved. The reader is referred to Coxeter [4] and Coxeter & Moser [6] for further information on the topics of this chapter. The latter book is difficult, partly because it is a survey, but the former book is very accessible.

For a general discussion of symmetry, including a discussion of tessellations and Escher's work, at a very elementary level, the reader is referred to Lockwood & Macmillan [21].

The Dutch artist M. C. Escher created many tessellations involving a variety of objects other than simple polygons as the fundamental region, some drawn on the hyperbolic plane represented by the unit disc. Any reader not familiar with this work is recommended to look at *The Graphic Work of M. C. Escher*, Oldbourne Book Co. Ltd, Pan/Ballentine 1961 [9].

8.6   EXERCISES

(1) What is wrong with the following proof that all triangles are isosceles?

Consider any triangle ABC. Construct the bisector of angle $\hat{A}$ and the perpendicular bisector DG of the side BC opposite to A. Let these bisectors intersect at D. Draw DE and DF perpendicular to AB and AC respectively, as shown in Fig. 8.6.1. Join D to B and D to C. Then triangles AED and AFD are congruent. Hence $AE = AF$ and $DE = DF$. Triangles BDG and DGC are congruent. Hence $BD = CD$. Triangles BED and CFD are congruent, since $DE = DF$, $BD = CD$, and $\hat{E} = \hat{F} = 90$ degrees. Thus $BE = CF$. Hence $AB = AE + EB = AF + FC = AC$. Therefore triangle ABC is isosceles. We conclude that all triangles are isosceles!

(2) If r_1, r_2, r_3, are reflections in lines in the Euclidean plane
   (a) show that $(r_1 r_2 r_3)^2$ is a translation;
   (b) what is the effect of $(r_2 r_3 r_1 r_2 r_3)^2$?

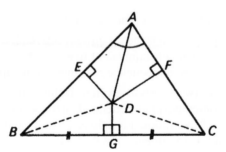

Fig. 8.6.1.

(3) Show that in the multiplication table of a group each element appears once and only once in each row and in each column.

(4) Using the result of exercise (3) above, together with the fact that if $ab = e$ in a group, then also $ba = e$, find all groups of orders 2, 3, and 4. (*Hint*: try to construct all possible multiplication tables.)

(5) Let D_3 be the set of symmetries of (transformations fixing) the equilateral triangle shown in Fig. 8.2.1. Let e denote the identity, u_1 reflection in the line L, u_2 reflection in the line M, u_3 reflection in the line N, and r rotation anti-clockwise through $2\pi/3$ radians. (Recall that lines L, M, N remain fixed in space throughout.) Taking the product $f \circ g$ of two symmetries f and g to mean the symmetry g followed by the symmetry f, find the multiplication table for D_3 and show that D_3 is a group. (It is a particular example of a *dihedral* group.)

Find the orders of the elements.

(6) Let $|G|$ denote the number of elements in the group G. (This is usually called the **order** of G.) Let H be a subgroup of the *finite* group G. Lagrange's theorem says that $|H|$ divides $|G|$. Use this fact to show that if $|G|$ is a *prime* number, then G must be *cyclic*, i.e. generated by óne element.

(7) Find *all* the subgroups of D_3.

(8) In a set W, if a, $b \in W$ are related according to some criterion R, say, we write $a R b$ and we say that R is an **equivalence** relation on the set W if

    (i) $a R a$ for *all* $a \in W$,

    (ii) whenever $a R b$, then also $b R a$,

(iii) whenever *a R b and b R c*, then also *a R c*.

An equivalence relation partitions the set into subsets, called **equivalence classes**, each consisting of all those elements equivalent to each other.

Verify that the following are equivalence relations on the appropriate sets.

(a) The equivalence of two *words* as given in Definition 8.2.4.

(b) The relation 'is homeomorphic to' defined on the set of all compact surfaces.

(c) The equivalence of two rational fractions a/b and c/d if and only if ad = bc, where a, b, c, and d are integers; $b \neq 0$, $d \neq 0$.

(d) The relation 'is isomorphic to' defined on the set of all graphs.

With the product defined as in Definition 8.2.5, verify that $[u][v] = [u'][v']$ when $u = x_1 x_2 x_3 x_4 x_4^{-1}$, $u' = x_1 x_2 x_2 x_2^{-1} x_3$ are equivalent *words* and $v = x_3 x_1^{-1} x_1 x_5$, $v' = x_2 x_2^{-1} x_3 x_5$ are equivalent *words*. That this must be checked in the *general* case is the content of the note following Definition 8.2.5.)

(9) On the Poincaré disc model of the hyperbolic plane draw a diameter and indicate the set of hyperbolic lines parallel to this line.

(10) In hyperbolic geometry, is parallel the same as equidistant? (Distance, of course, means *hyperbolic distance*.) In other words, given parallel lines *l*, *m* and points *A*, *B* on *l*, if we drop perpendiculars *AC* and *BD* onto *m*, is it necessarily true that *AC* and *BD* have the same hyperbolic length?

(11) How would you find the reflection of a point *p* in a line in the hyperbolic plane using compasses only?

(12) Using compasses only, and the idea of inversion in a circle, describe a procedure for dividing a line segment in the Euclidean plane into n equal parts.

(13) Sketch the reflection of the vector field for a source, shown in Fig. 8.6.2, in the line *C* in the hyperbolic plane.

(14) (a) By using Theorem 8.3.2, taking $K = 1$, show that the area of an n-gon in the hyperbolic plane is equal to its angular defect (i.e. the amount by which its angle sum falls short of that of an n-gon in the Euclidean plane).

(b) Let *ABC* be the 'asymptotic triangle' shown in Fig. 8.6.3, and suppose that the line *AD* bisects the angle $\hat{A}$. If B', E' are the reflections of *B*, *E* respectively, in *AD*, show that the area of $ABEE'B'A$ is equal to the area of *ABC*.

(15) Show that, for any polygon with angle sum $2\pi/n$, the half turns through the mid-points of the sides of the polygon produce a tessellation in the plane.

(16) Sketch tessellations, and their duals, for

(a) 2*K*;   (b) 3*T*;   (c) 5*P*.

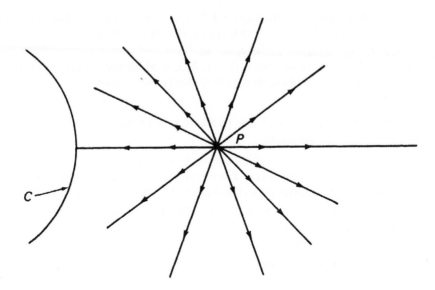

Fig. 8.6.2.

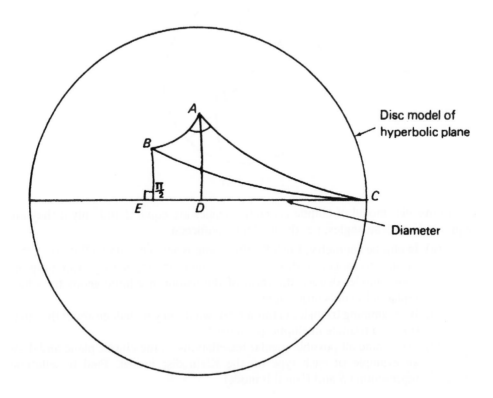

Fig. 8.6.3.

(17) (a) Write down the isometries a_i ($i = 1, 2, 3, 4, 5, 6$) for the tessellation $\{12, 12\}$ representing $3T$ and check that they satisfy the relation
$$a_1 a_2 a_1^{-1} a_2^{-1} a_3 a_4 a_3^{-1} a_4^{-1} a_5 a_6 a_5^{-1} a_6^{-1} = e.$$

   (b) Write down the isometries a_i ($i = 1, 2, 3, 4, 5$) for the tessellation $\{10, 10\}$ representing $5P$ and check that they satisfy the relation
$$a_1^2 a_2^2 a_3^2 a_4^2 a_5^2 = e.$$

(18) The following is a *disc model*, due to Klein, of what we call *elliptic geometry*. In this geometry, the **points** are

   (i) the Euclidean points forming the *interior* of a disc, and
   (ii) pairs of antipodal points i.e. Euclidean points on the boundary of the disc which are at opposite ends of a diameter represent a single elliptic point.

(*Notice* that the points of this model are essentially the points of the projective plane, as represented by a plane model.)

The elliptic **lines** are either

   (i) diameters of the disc, or
   (ii) arcs of circles meeting the boundary of the model at the ends of a diameter.

Two lines in this model of elliptic geometry are shown in Fig. 8.6.4.

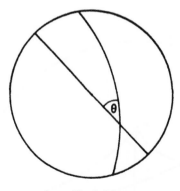

Fig. 8.6.4.

In this disc model of elliptic geometry, angles are equal if and only if they are equal as Euclidean angles, i.e. the model is **conformal**.

   (a) In elliptic geometry, Euclid's fifth axiom is again negated. (But now other axioms no longer hold, e.g. lines are finite in length.) By examining the above model deduce the form of the axiom in elliptic geometry which replaces Euclid's fifth axiom.
   (b) By examining triangles in this model, what can you deduce about the angle sum of a triangle in elliptic geometry?
   (c) Determine all possible regular tessellations on the elliptic plane and draw an example of each type on the Klein disc model. Find tessellations representing S and P on this model.

9

Some applications of tessellation representations

9.1 INTRODUCTION

In the previous chapter, we showed how to obtain plane tessellation representations of any compact surface, other than the sphere S and projective plane P, on either a Euclidean or a hyperbolic plane. In each of these representations, the faces of the tessellation are plane models of the compact surface represented. Thus, without actually constructing the space model, we can see how the edges of the plane model fit together to form the space model. In this chapter we exploit this fact to find when a pattern is a complex, to colour maps, and to examine vector fields on a given compact surface.

9.2 TESSELLATIONS AND PATTERNS

It is not always obvious from a plane model of a compact surface whether a pattern drawn on it is a complex or not. This is especially so if some of the vertices of the pattern lie on the edges of the plane model, or if some of the faces enclosed by the polygons of the pattern 'pass over' the edges of the plane model. By drawing the pattern on the faces of a tessellation representation of the compact surface, we can see all the faces of the pattern as complete polygons, which makes it much easier to see if the pattern is a complex.

In practice, since each face of the tessellation is a plane model of the compact surface and contains the complete pattern, we need concern ourselves only with a single face of the tessellation, and with those faces of the pattern which are *not* entirely enclosed within this face of the tessellation. Each distinct face of the pattern in this chosen face of the tessellation can be numbered, and the adjoining faces of the pattern in the neighbouring faces of the tessellation can then be numbered accordingly. By examining each face of the pattern in turn, we can see whether or not the conditions for a complex are satisfied.

A convenient method for checking whether a pattern is a complex is the following.

First check the more obvious things. Is each face a genuine polygon, homeomorphic to a disc and contractible to a point on the surface? Then, to check that faces meet correctly, use the following theorem.

THEOREM 9.2.1   Let D be a mosaic of polygons covering the compact surface M. Let all the polygons in D have at least three edges, and let all the vertices in D lie on at least three edges. Let h be the shortest distance between two vertices of the mosaic D on the tessellation representing M. Draw a line l_x around the face of D labelled x and at a distance h/2 from the edges of face x on the tessellation. The D is a complex if and only if for *each* face x

> (a) the line l_x does *not* pass through face x, and
> (b) the line l_x does *not* pass through any face of D more than once.

PROOF   It is not difficult to see that a face x will fail to meet itself, and will meet every other face in D in the manner required in a complex, if and only if the loop l_x satisfies the conditions (a) and (b). The detailed proof of this requires the consideration of several cases and is outlined as an exercise at the end of the chapter.

We now look at two examples illustrating the use of the method provided by Theorem 9.2.1.

EXAMPLE 9.2.1   Consider Ungar's pattern on a torus as shown in Fig. 9.2.1. We previously gave this pattern in a slightly different form in Fig. 6.1.3.

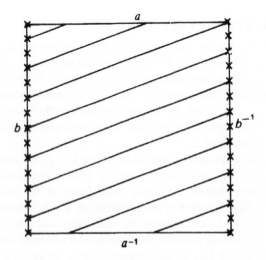

Fig. 9.2.1.

Ungar's pattern was designed to demonstrate that there is at least one pattern on the torus which needs seven different colours if each face of the pattern is to be coloured and no two faces of the same colour are to meet in an edge. It is not at all obvious that this is so from a *single* plane model of the torus. However, once the pattern is superimposed on the tessellation representing the torus, it becomes clear where the vertices and edges of the pattern meet. This is shown in Fig. 9.2.2.

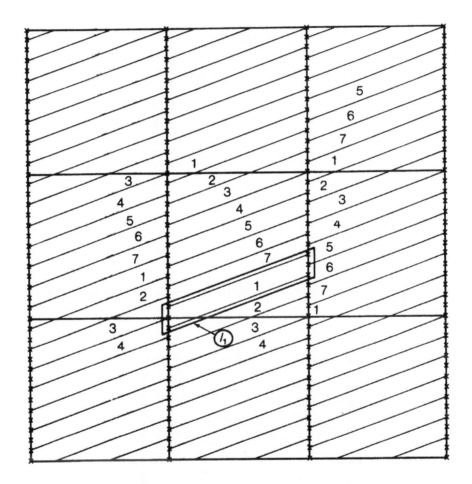

Fig. 9.2.2 — Ungar's pattern on the tessellation representing the torus T.

To find out if this pattern is a complex, and therefore also a map, we draw the line l_x around each face x of the pattern. Then, by noting down each face through which the line passes, moving clockwise around the face x, we obtain a sequence of labels, usually natural numbers, associated with the face x. We call this sequence a **reading** for the face x.

By Theorem 9.2.1, assuming that each polygon and vertex of the pattern has at least three edges, if, for each face x in the pattern the reading for x does *not* contain x or *two* instances of y, where y is *any* other face of the pattern, then the pattern is a complex.

According to the numbering of the faces given in Fig. 9.2.2, Ungar's pattern on the torus gives the readings of Table 9.2.1. Hence, Ungar's pattern is a complex on the torus.

Table 9.2.1

Face	Reading					
1	7	5	6	2	4	3
2	1	6	7	3	5	4
3	2	7	1	4	6	5
4	3	1	2	5	7	6
5	4	2	3	6	1	7
6	5	3	4	7	2	1
7	6	4	5	1	3	2

EXAMPLE 9.2.2 Now consider a similar pattern on the Klein bottle, as shown in Fig. 9.2.3.

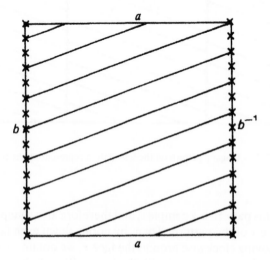

Fig. 9.2.3.

Fig. 9.2.4 shows Ungar's pattern drawn on the tessellation representation of K. The pattern is drawn on the fundamental region of the tessellation and then is transformed with it to obtain the other faces of the tessellation. Thus, the pattern is 'flipped over' in certain of the faces of the tessellation. This is apparent in Fig. 9.2.4.

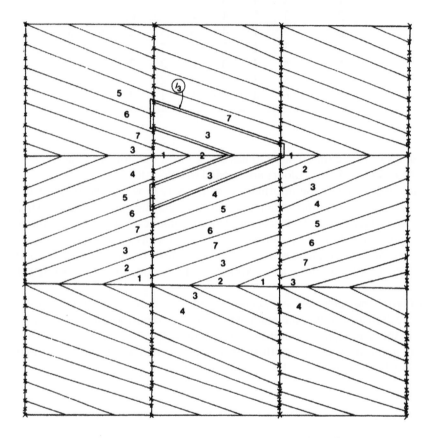

Fig. 9.2.4 — Ungar's pattern on the tessellation representing the Klein bottle K.

According to the numbering of the faces of the pattern given in Fig. 9.2.4, the readings are as given in Table 9.2.2. This time the pattern is *not* a complex, since face 3 of the pattern meets face 5 in two edges.

9.3 TESSELLATIONS AND MAP COLOURING

Ungar's pattern on a torus is designed so that every face meets each of the other six faces. This information is carried in Table 9.2.1. It is obvious that exactly seven colours are needed to colour this map.

We have seen that Ungar's pattern on a Klein bottle, as shown in Fig. 9.2.3, is not a complex. However, this pattern is a *map*, according to Definition 6.1.1, and Table

Table 9.2.2

Face								
1	2	4	3	7				
2	3	5	4	1	7	6		
3	7	1	4	6	**5**	2	6	**5**
4	3	1	2	5	7	6		
5	4	2	**3**	6	**3**	7		
6	5	3	4	7	2	3		
7	6	4	5	3	1	2		

9.2.2 helps us to see this. This map has seven faces, but not every face meets every other face, so how many colours are needed to colour it?

Table 9.2.2 carries the information on the way in which the faces of this map meet. We can use this information in order to colour the map in the following way.

Choose a face x and let it be coloured a_1. This implies that the faces in the reading of x may *not* be coloured a_1. Suppose that i colours have already been assigned. If there is an uncoloured face y which cannot be coloured by any of the i colours already used, then let it be coloured a_{i+1}. *No* face in the reading of y can now be coloured a_{i+1}.

If there is no such face as y, choose some uncoloured face z and some colour $a_j, j \leq i$, such that z is not in the reading for *any* face coloured a_j. Let z be coloured a_j. Faces in the reading of z cannot then be coloured a_j.

One application of this method to the pattern under consideration is given in Table 9.3.1. In the table, let face 1 be coloured a_1.

Table 9.3.1

Face	1	2	3	4	5	6	7
Colours	a_1	$\sim a_1$	$\sim a_1$	$\sim a_1$			$\sim a_1$
		a_2	$\sim a_2$	$\sim a_2$	$\sim a_2$	$\sim a_2$	$\sim a_2$
			a_3	$\sim a_3$	$\sim a_3$	$\sim a_3$	$\sim a_3$
				a_4	$\sim a_4$	$\sim a_4$	$\sim a_4$
					$\sim a_5$	$\sim a_5$	a_5
					a_1	$\sim a_1$	
						a_6	

The symbol $\sim$ means not.

Face 1 meets face 2, so let 2 be coloured a_2.

Face 3 meets an a_1 colour and an a_2 colour, so let 3 be coloured a_3.

Face 4 meets colours a_1, a_2, and a_3, so let 4 be coloured a_4.

Face 7 meets colours a_1, a_2, a_3, a_4, so we colour 7 with a_5.

Face 5 meets colours a_2, a_3, a_4, a_5, but not a_1, hence we colour 5 with a_1.
There is no need to introduce a new colour.

Face 6 meets all five colours used so far, so we colour 6 with a_6.

Thus Ungar's pattern on the Klein bottle K can be coloured with six colours. It has been shown that six colours are sufficient to colour all maps on the Klein bottle, and we have verified here that this number cannot be reduced. The Klein bottle provides the one exception to Heawoods's conjecture; see the end of section 6.1.

When colouring maps, we are interested only in pairs of faces which meet along complete edges. Pairs of faces which meet at a single vertex may be given the same colour. The example just considered contained no such pair of faces. To apply the above method to colour any given map on any given compact surface, we must introduce the following modification of the idea of a reading.

If we delete from the reading for x each face which, at that position, meets x only at a vertex, we get what we shall call an **edge reading** for x. Thus the edge reading for x contains the faces meeting x along a complete edge. As before, we move in a clockwise sense round the face x. Our previous technique can now be applied, using the table of edge readings of the map, in order to colour any given map on a compact surface.

EXAMPLE 9.3.1   Consider the pattern on $3P$, the connected sum of 3 projective planes, shown in Fig. 9.3.1.

Fig. 9.3.2 shows this pattern superimposed on the tessellation representation of $3P$ obtained as in Example 8.4.4 (with n = 3).

Moving round the line l_9 shown in Fig. 9.3.2, in a clockwise sense, we find that the reading for the face 9 is as given in Table 9.3.2. Thus the face 9 meets itself and also meets faces 5 and 12 twice. Hence the pattern is not a complex on $3P$. However, we can see from Fig. 9.3.2 that it is a map. The *edge* readings for this map are as given in Table 9.3.3.

It is left as an exercise to produce a colouring for the map using the edge readings given in Table 9.3.3. Using an argument similar to that used with Table 9.3.1, the reader should find that no more than five colours are needed to colour this map.

9.4 TESSELLATIONS AND VECTOR FIELDS

As with patterns, tessellation representations of compact surfaces can be used to give a more complete picture of a vector field on a compact surface than can be obtained from the plane model alone. Essentially, by superimposing the vector field on the tessellation represention, we can see how the vector field at the edges of the plane model joins together, without having to construct a space model.

In Figs 9.4.1 and 9.4.2, we give straightforward examples of vector fields without critical points on tessellations representing the torus T and the Klein bottle K respectively. The tessellations used are those given in Examples 8.4.1 and 8.4.2.

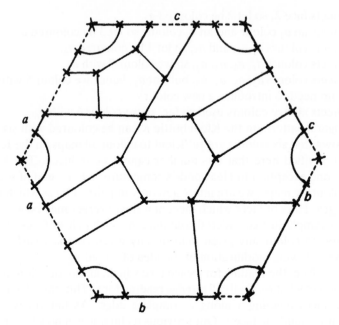

Fig. 9.3.1.

Again, we develop our ideas through a sequence of examples. Usually we shall be concerned with the special kind of vector field called a **direction field**; see Definition 7.2.2. (Some authors call this a **tangent field**).

EXAMPLE 9.4.1   Consider the direction field on the compact surface $3P$ shown in Fig. 9.4.3.

From this plane model, it is not at all clear where the critical points are, or what their indices are. However, the situation is clarified when the direction field is superimposed on the tessellation represention of $3P$ which we used in Example 9.3.1. This is shown in Fig. 9.4.4.

From Fig. 9.4.4, it can be seen that there is one critical point at the vertex of each face of the tessellation. (Recall that each face of the tessellation is a plane model of $3P$.)

Now each vertex of the plane model represents the same single point on the space model of $3P$. thus, altogether, there is *one* critical point on the compact surface $3P$.

The arrows around each critical vertex make one complete revolution clockwise when we travel around the critical point in an anti-clockwise sense. Thus, by Definition 7.2.3, the index at the critical point on $3P$ is -1. Since this is the only critical point, the sum of the indices is -1, which is also the Euler characteristic of $3P$. Thus, although $3P$ is *non*-orientable, this agrees with the index Theroem 7.2.1, which we stated only for *orientable* compact surfaces.

EXAMPLE 9.4.2   In the direction field on $2T$ shown in Fig. 9.4.5, the critical points lying within the plane model are easy to see, and those lying on the edges but not at

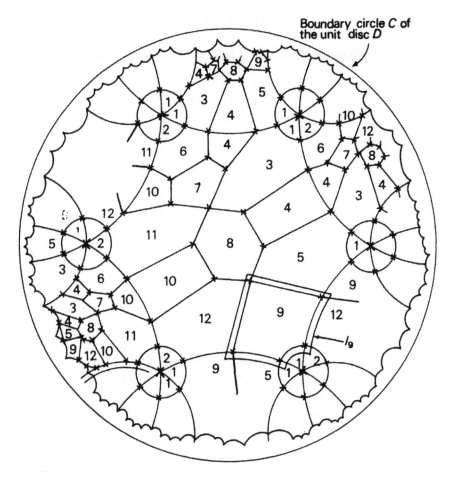

Fig. 9.3.2 — The pattern of Fig. 9.3.1 superimposed on the tessellation representation of 3P on the hyperbolic plane.

Table 9.3.2

Face		Reading							
9	5	9	12	2	1	5	9	12	8

vertices are quite easy to guess. However, it is not so easy to spot the critical point arising from the vertices of the plane model. This becomes clear when the direction field is superimposed on the tessellation representation of $2T$ on the hyperbolic plane given in Example 8.4.3.

Table 9.3.3

Face	Edge reading						
1	3	5	9	2			
2	6	1	12	11			
3	6	4	8	7	4	5	1
4	7	6	3	8	5	3	
5	9	9	8	4	3	1	
6	7	10	11	2	3	4	
7	11	10	6	4	3		
8	5	12	10	11	3	4	
9	12	1	5	12	5		
10	12	11	7	6	11	8	
11	10	6	2	12	10	7	8
12	9	2	11	10	8	9	

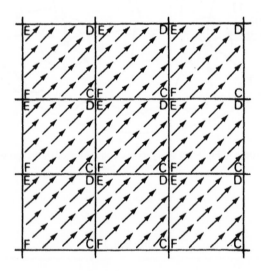

Fig. 9.4.1 — A vector field with no critical points on the tessellation representing the torus T.

It is convenient to take a vertex of the (dual) tessellation at the centre of the hyperbolic plane as shown in Fig. 8.4.21. This produces Fig. 9.4.6.

From Fig. 9.4.6, we can see that the central critical point has eight hyperbolic sectors and so, using formula 7.2.1, i.e. index $= 1 + (a - c)/2$, we see that the index of this critical point is -3.

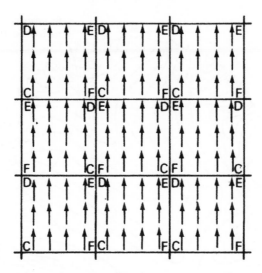

Fig. 9.4.2 — A vector field with no critical points on the tessellation representing the Klein bottle *K*.

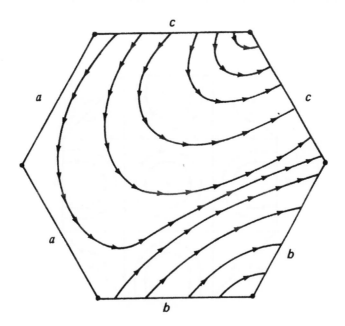

Fig. 9.4.3 — A direction field on the compact surface 3*P*.

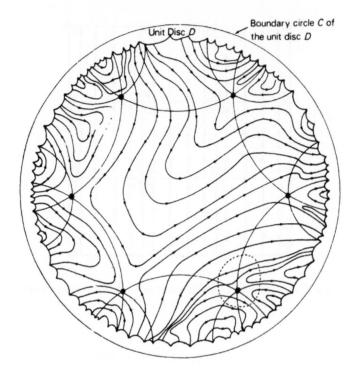

Fig. 9.4.4 — The vector field of Fig. 9.4.3 superimposed on a tessellation representation of $3P$ on the hyperbolic plane. A broken circle surrounds a typical critical point.

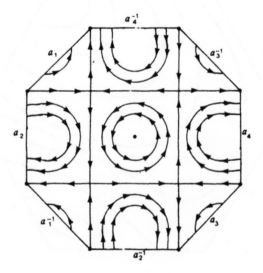

Fig. 9.4.5 — Direction field on $2T$.

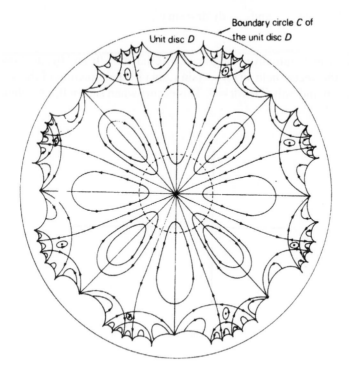

Fig. 9.4.6 — The direction field of Fig. 9.4.5 superimposed on a tessellation representation of
2*T* on the hyperbolic plane. A broken circle surrounds the central critical point.

The eight edges meeting the central vertex in Fig. 9.4.6 represent four lines on the
space model of 2*T*. Thus the eight critical points lying on these edges represent four
critical points on the space model of 2*T*. Each of these latter critical points has index
1.

Within each face of the tessellation, i.e. inside the plane model of 2*T*, there are
four crosspoints, each with index − 1, and one centre with index 1.

Hence the sum of the indices is

$$-3 + 4(1) + 4(-1) + 1 = -2 ,$$

which is the Euler characteristic of the double torus in agreement with the index
Theorem 7.2.1.

As pointed out in section 7.7, vector fields and differential equations are closely
related. The following example shows how the integral curves of some differential
equations can be naturally associated with specific compact surfaces.

EXAMPLE 9.4.3   We consider the direction field associated with the differential
equations

$$dx/dt = \sin x \quad \text{and} \quad dy/dt = \sin y ,$$

relative to the usual mutually orthogonal Cartesian axes $0x$ and $0y$, as shown in Fig. 9.4.7. This is the vector field $V(x,y) = (\sin x, \sin y)$. The direction of the vector field will be *vertical* if and only if $dx/dt = 0$. This occurs when $\sin x = 0$; i.e. when $x = m\pi$, $m = 0, \pm 1, \pm 2, \ldots$.

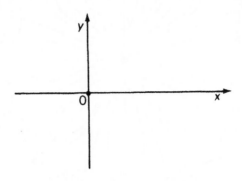

Fig. 9.4.7.

When $2m\pi < y < \pi + 2m\pi$, $\sin y$ is positive, hence the arrows point 'upwards'.

When $(2m + 1)\pi < y < 2(m + 1)\pi$, $\sin y$ is negative, hence the arrows point 'downwards'.

The arrows of the vector field will be *horizontal* if and only if $dy/dt = 0$. This occurs when $\sin y = 0$; i.e. $y = m\pi$, $m = 0, \pm 1, \pm 2, \ldots$.

When $2m\pi < x < (2m + 1)\pi$, $\sin x$ is positive, hence the arrows point towards the right.

When $(2m + 1)\pi < x < 2(m + 1)\pi$, $\sin x$ is negative and the arrows point towards the left.

From this, it follows that the vector, or direction, field is doubly periodic, along the two axes $0x$ and $0y$, with period 2π.

By superimposing a grid, with cells $2\pi \times 2\pi$, over the direction field, as shown in Fig. 9.4.8, we can see that the translations used to generate the tessellation representation of the torus T in Example 8.4.1 map the direction field *onto itself*.

Hence the direction field fits into the tessellation representation of the torus with each square of the grid as a face of the tessellation. This means that it is possible to represent the direction field

$$V(x, y) = (\sin x, \sin y)$$

on a torus.

In Fig. 9.4.9, we show the direction field drawn on a space model of the torus T.

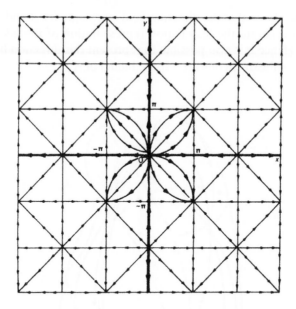

Fig. 9.4.8 — The direction field $V(x, y) = (\sin x, \sin y)$ superimposed on a $2\pi \times 2\pi$ grid. Arrows are vertical on vertical lines $\sin x = 0$ and horizontal on the horizontal lines $\sin y = 0$.

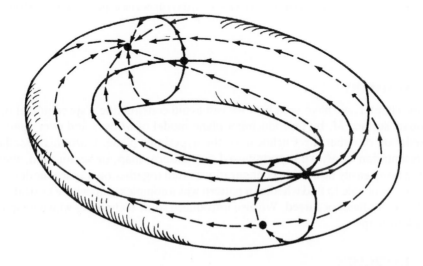

Fig. 9.4.9 — Direction field $V(x, y) = (\sin x, \sin y)$ drawn on a space model of a torus T.

Fig. 9.4.8 also shows that the direction field maps onto itself under the transformations used to generate the tessellation representation of the Klein bottle K in Example 8.4.2. Hence it is also possible to represent this direction field on a Klien bottle. Fig. 9.4.10 illustrates this for a space model of K.

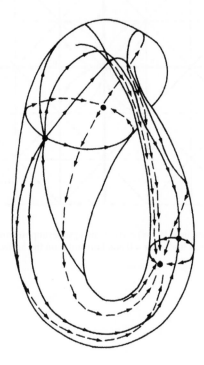

Fig. 9.4.10 — The direction field $V(x, y) = (\sin x, \sin y)$ drawn on a space model of a Klein bottle K.

9.5 SUMMARY

In this chapter, we have seen how we can avoid constructing a space model of our compact surface M, by reproducing a plane model of M over and over again in a tessellation of either the Euclidean or the hyperbolic plane. Onto this tessellation representation of M, we superimposed our pattern, map, or vector field, thereby seeing fairly easily how these configurations fitted together on a space model. From this, we were able to find whether a pattern was a complex, and where critical points of the vector fields occurred. We also introduced some techniques which enabled us to colour maps.

9.6 EXERCISES

(1)(To prove Theorem 9.2.1.) Show

(A) if l_x passes through the face x, then x meets itself,

(B) if l_x passes through the face y *twice*, then *either* (i) the face y meets x twice at a single vertex, and so y meets itself, *or* (ii) the face y meets x in one of three other possible ways, and, in each case, this either implies that x meets itself or it implies that x fails to meet y correctly.

To prove the converse, show that if the mosaic D fails to be a complex, then one of the conditions (a) or (b) in Theorem 9.2.1 must fail.

(2) Show that if a mosaic D on a compact surface M has n faces, then

(a) if D is a pattern, the number of entries in its reading is even,

(b) if D is a complex, the number of entries in its reading is $\leqslant n(n-1)$.

(3) (a) Using tessellations, repeat Exercise 5.6.1(d) and (f).

(b) Find colourings of the maps using just 4 and 5 colours respectively.

(4) Using Table 9.3.3, find a colouring of the map in Fig. 9.3.1 using just five colours.

(5) Colour your map of an icosahedron, produced in Exercise 5.6.4(b), with three colours.

(6) In how many essentially different ways can

(a) a cube be coloured with three colours,

(b) a dodecahedron be coloured with four colours?

(7) Colour the vertices of the graph in Fig. 6.9.2 with a minimum number of colours, so that no two vertices of the same colour are linked by an edge.

(8) Lable the vertices of the graph in Fig. 9.6.1 with the letters a, b, c, d, e, f, g, h so that no pair of vertices linked by an edge are labelled with adjacent letters.

(9) Sketch the separatrixes for the critical point in Fig. 9.4.3 and check its index using the formula $1 + (a - c)/2$.

(10) Sketch the dual of the vector field shown in Fig. 9.4.5 and sum its indices.

(11) Determine the sum of the indices of the critical points in the vector field $V(x, y) = (\sin x, \sin y)$ shown in Figs 9.4.8 and 9.4.9, comparing your solution with the result expected by the index theorem.

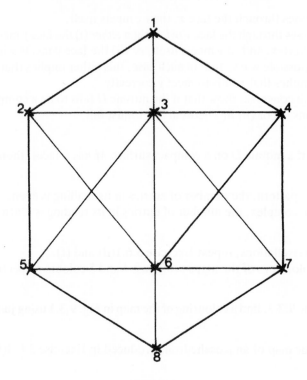

Fig. 9.6.1.

(12) Sketch the vector field $V(x, y) = (\sin x - \sin y, -\sin y)$ and superimpose a tessellation on it, in order to show that it fits naturally onto a torus T.

It is possible to fit a tessellation representation of the Klein bottle K onto this
vector field?

10

Introducing the fundamental group

10.1 INTRODUCTION

In the course of discussing Example 8.4.1, we mentioned the fundamental group. In the standard treatment of surface topology, after developing a background of general topology based on set theory, the compact surface is described algebraically in terms of the fundamental group. To make our approach more generally accessible, we have avoided this by describing a compact surface as a 2n-gon and representing it by the associated *word*. This procedure is described in detail in section 3.5. In our treatment the *word* replaces the fundamental group, which is, therefore, not needed.

However, in order to make it easier for the reader to make the transition from the development in this book to the standard development, we give a definition of the fundamental group in terms of our treatment. First we mention in outline how the fundamental group is usually defined.

We take a torus to illustrate the ideas involved, as shown in Fig. 10.1.1

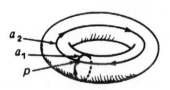

Fig. 10.1.1 — Space model of a torus T.

Take some fixed point p on the torus. Draw loops a_1 and a_2 so that each loop starts and finishes at the point p, called the base point. Two loops starting and finishing at p are considered to be *equivalent* (essentially the same) if one can be continuously deformed into the other within the surface.

The product of two loops is the path on the torus obtained by first going round one loop and then going round the other loop. The inverse of a loop is a loop traversed in a direction opposite to that of the former. On the torus there are essentially two distinct loops, and their inverses, from which all paths from p to p, up to equivalence, can be obtained. These are the loops a_1 and a_2 shown in Fig. 10.1.1. The set of all paths from p to p (up to equivalence), with the product of two paths defined as one path followed by the other, is a group. This group, generated by a_1 and a_2, is called the **fundamental group** of the compact surface T, and is written $\pi(T)$. It can be shown to be independent of the choice of the base point p. Fig. 10.1.2 shows the element $a_1^3 a_2$ in $\pi(T)$. Here we mean the path obtained by going round a_2 once in the direction of the arrow and then going three times round a_1 in the direction of the arrow.

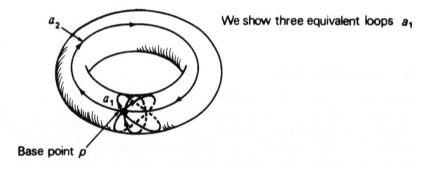

Fig. 10.1.2 — Showing the element $a_1^3 a_2$ of $\pi(T)$.

For the compact surfaces M dealt with in this book $\pi(M)$ characterizes M. Unfortunately, it is not always easy to calculate $\pi(M)$, which is another reason for the difficulty of the standard approach.

10.2   THE FUNDAMENTAL GROUP

In this section, we give our definition of the fundamental group based on our characterization of compact surfaces by *words*. Obviously we are unable to prove that our definition is equivalent to the standard one, since this would involve going into the standard approach more deeply than is appropriate for our purpose. However, as you can see in Fig. 10.1.1, there is an obvious connection between the *loop* a_i ($i = 1, 2$), which forms an element of the group, and the *edge* a_i in our plane model, which was formed by slicing the space model along the loops a_1 and a_2: see section 2.1.

Let M be *any* compact surface. By Theorem 3.6.1, the classification theorem, M can be taken to be one of the following:

$S, P, K, nT, (nT)P,$ or $(nT)K$, for some $n \geq 1$.

Corresponding to these 2n-gons, we have the *normal word* representations:

$$aa^{-1}, aa, a_1 a_2 a_1^{-1} a_2 ,$$
$$a_1 a_2 a_1^{-1} a_2^{-1} . a_3 a_4 a_3^{-1} a_4^{-1} \ldots a_{2n-1} a_{2n} a_{2n-1}^{-1} a_{2n}^{-1} ,$$
$$a_1 a_2 a_1^{-1} a_2^{-1} \ldots a_{2n-1} a_{2n} a_{2n-1}^{-1} a_{2n}^{-1} . a_{2n+1} a_{2n+1} ,$$
$$a_1 a_2 a_1^{-1} a_2^{-1} \ldots a_{2n-1} a_{2n} a_{2n-1}^{-1} a_{2n}^{-1} . a_{2n+1} a_{2n+2} a_{2n+1}^{-1} a_{2n+2} ,$$

as discussed in section 3.5

We can now give

DEFINITION 10.2.1   Let M be any compact surface with *normal word* representation W as given above. Let the symbols involved in W be $a_1, a_2, a_3, \ldots, a_t$. Then the group

$$\langle a_1, a_2, \ldots, a_t | W = e \rangle ,$$

generated by the symbols $a_1, a_2, \ldots, a_t$ subject to the relation $W = e$, is called the **fundamental group** of M, written $\pi(M)$.

Notice that once we know W, we can at once write down $\pi(M)$.

Now suppose that we have a *word* representation W' of the compact surface M, where W' is *not* the normal *word* W. Consider the group

$$F = \langle b_1, b_2, \ldots, b_s | W' = e \rangle ,$$

where $b_1, b_2, \ldots, b_s$ are the symbols occurring in W'.

What relation does this group F have to the fundamental group $\pi(M)$?

To answer this question, it is convenient to have available the notion of *isomorphic* groups.

10.3   ISOMORPHIC GROUPS

In section 8.2 we discussed the group

$$G = \langle x_1, x_2 | x_1^2 = x_2^2 = (x_1 x_2)^2 = e \rangle .$$

Let us now consider the group

$$H = \langle y_1, y_2 | y_1^2 = y_2^2 = e, \, y_1 y_2 = y_2 y_1 \rangle$$

Since $y_1 y_2 = y_2 y_1$, the group H must be abelian. A little consideration shows that the elements of H are:

$$e, y_1, y_2, \text{ and } y_1 y_2 ,$$

and that the multiplication (Cayley) table loooks exactly like the one for G which we obtained in section 8.2. The only difference is in the symbols used. Groups like this, that differ only in how we write the elements, are called **isomorphic** groups. The reader is invited to check that the following two groups are isomorphic.

$$\langle a, b \,|\, a^2 = b^2 = (ab)^3 = e \rangle$$

and

$$\langle x, y \,|\, x^3 = y^2 = e, \; xy = yx^2 \rangle$$

One way of doing this is to find the multiplication tables of the two groups and then try to make the elements of the groups correspond in such a way that the two tables are manifestly the same. The one–one correspondence set up in this way between isomorphic groups is usually called an **isomorphism**.

Another way of showing that two groups expressed in this way are isomorphic is to manipulate the generators and relations defining one group until they have the same form as for the other group; such manipulations are provided by the so-called **Tietze transformations**. Certain rules of manipulation must hold and the process can be very difficult to carry out. In fact, there is *no general* procedure which will always work.

We are faced with this situation with the groups F and $\pi(M)$. However, it turns out that by using Operations 1–3 of section 3.3 and Operation 4 of section 3.4, we can change one group into the other. This shows that the groups F and $\pi(M)$ are *isomorphic*. A consequence of this is that it does not matter what *word* representation we choose for M in order to find $\pi(M)$. The *word* representation that we have chosen in Definition 10.2.1 gives what we might call the **normal presentation** of $\pi(M)$.

Another question that arises naturally is the following.

If $\pi(M_1)$ is isomorphic to $\pi(M_2)$, how are the compact surfaces M_1 and M_2 related?

It can be shown that if $\pi(M_1)$ is isomorphic to $\pi(M_2)$, then the *word* W_1 defining $\pi(M_1)$ can be transformed into the *word* W_2 defining $\pi(M_2)$ by using Operations 1–4 of sections 3.3 and 3.4. But, by Definition 3.5.2, this means that M_1 and M_2 are *homeomorphic*, i.e. from the topological point of view, they are essentially the same.

Altogether, $\pi(M)$ uniquely characterizes the compact surface M. For us it is a more complicated way of achieving this than the *word* representation, but, as described in the next section, it comes into its own for higher dimensions than two.

10.4 COMMENTS

We have been concerned in this book exclusively with spaces of dimension two, i.e. surfaces. In this situation our *word* representation is adequate, or, if we wish to develop the theory within the context of contemporary set theory, the fundamental group is adequate. The latter is also efficient for spaces of dimension three. However, this is no longer true for spaces M of dimension n, n > 3. In this latter case, one method of approach is to define a sequence of groups associated with the space M. We can do this by turning the set of all loops used for the fundamental group into a *new space M'* in its own right. Then the first group in the sequence is taken to be $\pi(M)$ and the second is taken to be $\pi(M')$. This process is repeated to produce the complete sequence of so-called **homotopy groups** $\pi_n(M)$. The first homotopy group $\pi_1(M)$ is the fundamental group. This approach forms the subject matter of that branch of algebraic topology known as **homotopy theory**.

Unfortunately, in general, the fundamental group is non-abelian, and non-abelian groups are very difficult to handle. Moreover the homotopy groups are described in terms of generators and relations. Usually, such a description of a group yields its secrets very reluctantly. For this reason an alternative approach offers its attractions.

Again a sequence of groups is associated with the space M, but this time an entirely different method of construction yields abelian groups. This approach is the subject matter of **homology theory**. From these beginnings, a branch of algebra, called **homological algebra**, has arisen in recent years. It has proved a tool of great power and wide application. In particular, apart from algebraic topology, it has been applied to group theory and number theory with considerable success.

Good introductions to this approach are given in Henle [16] and Giblin. [13].

10.5 EXERCISES

(1) Show that $G = \langle a, b \,|\, a^2 = b^2 = (ab)^3 = e \rangle$ and $H = \langle x, y \,|\, x^3 = y^2 = e, xy = yx^2 \rangle$ are isomorphic.

(2) Show that the groups G and H above are isomorphic to D_3. (See Exercise 8.6.5)

(3) Let G be an infinite cyclic group, i.e. a group $G = \langle g \rangle$ generated by the single element g of infinite order. Show that G is isomorphic to the group of integers $\mathbf{Z}$ under ordinary addition of integers.

(4) Show that $G = \{1, -1\}$ is a group under the ordinary multiplication of integers, and that $\pi(P) \cong G$. ($\cong$ denotes **'is isomorphic to'**.)

(5) Define a **'product'** on the set

$$\mathbf{Z} \times \mathbf{Z} = \{(a,b) \,|\, a, \; b \in \mathbf{Z}\}$$

by

$$(a_1, b_1) \,.\, (a_2, b_2) = (a_1 + a_2, b_1 + b_2) \ .$$

Show that $\mathbf{Z} \times \mathbf{Z}$ under this product is a group isomorphic to $\pi(T)$.

(6) In Exercise 1.11.9, you considered the idea of a 1-dimensional surface. Now, by taking a solid sphere as a basic 3-dimensional nighbourhood of a point, mimic the definitions in Chapter 1 in order to define a '3-dimensional manifold', a '3-dimensional surface', and a 'compact 3-dimensional surface'.

(a) To produce a 3-dimensional torus, we identify opposite faces of a cube. Show that under this procedure the edges are identified in sets of 4, while all 8 vertices are identified to a single point.

   Sketch the neighbourhoods for points on the faces, on the edges, and at the vertex of the 3-dimensional torus, checking that these neighbourhoods are equivalent to solid spheres.

(b) Consider the sets formed by identifying opposite faces of

   (i) a hexagonal prism,
   (ii) an octagonal prism.

   In each case, check the pattern of identification of edges and vertices, and check to see if the set is a 3-dimensional manifold.

(c) Consider the possibility of tessellation representations of the 3-dimensional manifolds in (a) and (b).

11

Surfaces with Boundaries

11.1 DEFINITIONS AND CLASSIFICATION THEOREMS

The sets studied so far in this book have been compact two-dimensional manifolds. Because they are compact any space model for the set in $\mathbb{R}^3$ or $\mathbb{R}^4$ must be closed and bounded. Because they are locally flat they can have no boundary points. Thus familiar sets like the disc have been excluded from our study. If we include the circular boundary the disc is compact but not locally flat. If we exclude the circular boundary the disc is locally flat but not closed, and therefore not compact.

All surfaces of real objects in space are compact two-dimensional manifolds so are covered by the approach so far taken. However, many sets we meet regularly in mathematics are of a type we would like to interpret as surfaces but which do not fit the compact two-dimensional manifold definition. These are sets which appear to be "surface-like" but which have one dimensional boundaries. So our problem in attempting to include such sets in our theory is to deal with these one dimensional boundaries. We do this by widening our idea of local flatness to include points on the boundary.

The sets we have already modelled as two-dimensional manifolds can be visualised as space models in $\mathbb{R}^n$ (for some $n \geq 3$), which we imagine to be made of a transparent locally flat film with no thickness - so when we mark a point on the "outside" of the film then view it from the "inside" we see the same point. The new sets we intend to model can be visualised in the same way but this time with some holes cut in the space model. In the definition we adopt the points on the one-dimensional boundaries, the boundaries of the holes, are included in the set and so the set is still compact.

We could summarise the definition of a compact two-dimensional manifold given in Chapter 1 as follows.

DEFINITION 11.1.1   The set S is a space model for a compact two-dimensional manifold, or a **surface without a boundary**, if

(1) S is locally flat,

(2) S is connected,

(3) S is homeomorphic to a compact subset of $\mathbb{R}^n$ for some $n \geq 3$.

Now the extension of this definition to include surfaces with holes is

DEFINITION 11.1.2 The set S is a space model for a **surface with boundary** if

(1) For each point P in S we can find

either

    (a) a homeomorphism from the open disc $\{(x, y) : x^2 + y^2 < 1\}$ to a neighbourhood of P, with the centre $(0, 0)$ of the disc mapped to P,

    or

    (b) a homeomorphism from the half-disc $\{(x, y) : x^2 + y^2 < 1, y \geq 0\}$ to a neighbourhood of P, with the point $(0, 0)$ in the half-disc mapped to P.

(2) S is connected.

(3) S is homeomorphic to a compact subset of $\mathbb{R}^n$.

This extension of the idea of local flatness to include points on a one-dimensional boundary is illustrated in Fig. 11.1.1 where we can see one of these new surfaces, a torus with a hole cut in it.

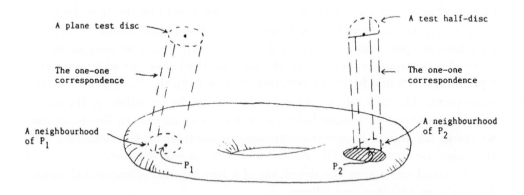

Fig.11.1.1

DEFINITION 11.1.3 The **boundary** of a surface with boundary is the subset of points P on the surface for which no homeomorphism exists of the type in condition 1(a), so that only a homeomorphism from a neighbourhood of P to a half-disc is possible, as in condition 1(b).

Now to extend our theory to include surfaces with boundaries we need to cut holes in our previously studied surfaces without boundaries.

Consider first of all a sphere with a hole in it. As can be seen by the sequence of distortions in Fig. 11.1.2 this surface with a boundary is home-omorphic to a closed disc

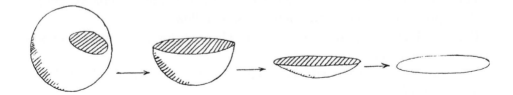

Fig.11.1.2

So let us denote this "sphere with a hole cut in it" by $\bar{D}$ (D stands for disc, and the bar indicates that the boundary is included.)

Now consider the connected sum of a sphere with a hole, $\bar{D}$, and a space model of some other surface M.

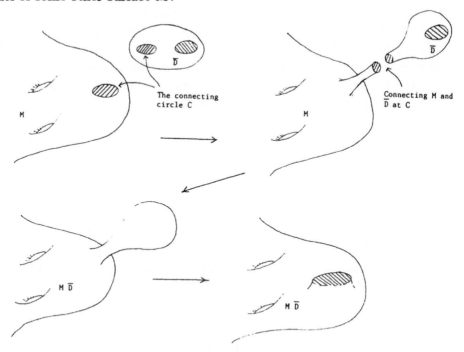

Fig.11.1.3

As shown in Fig. 11.1.3 the effect of this connected sum is to cut a hole in M. So $M\#\bar{D}$, or $M\bar{D}$ for short, is now a surface with boundary. To distinguish these holes with a one-dimensional boundary from the sort of hole we have through the centre of a torus, for example, let us call them

cuffs. Thus we have seen that one cuff cut in some surface M corresponds to $M\bar{D}$. Two cuffs cut in M will correspond to $M\#\bar{D}\#\bar{D}$, or $M(2\bar{D})$ for short, and so on.

Now in Theorem 2.4.1 we gave a complete description of surfaces without boundaries. By considering these surfaces and then adding cuffs we can get a description of surfaces with boundaries, as follows.

THEOREM 11.1.1 (The Classification Theorem for space models of surfaces with boundary).

An orientable surface with boundary is homeomorphic to $nT\#m\bar{D}$, for some $n \geq 0$ and $m \geq 1$.

A non-orientable surface with boundary is homeomorphic to

$$\text{either} \quad nT\#K\#m\bar{D}$$
$$\text{or} \quad nT\#P\#m\bar{D}$$

for some $n \geq 0$ and $m \geq 1$.

So far our discussion has been concerned with surfaces with boundary considered as sets in space, i.e. with space models of these surfaces. In order to work more comfortably with such sets our next task is to model them combinatorially. We have already done this for surfaces without boundary, so to extend this we need to model the cuffs combinatorially. In other words we need to transfer the idea of creating a cuff in a given space model into the language of plane models and their associated words. So we need to consider the plane model, and the word, for a disc.

Fig. 11.1.4. shows how $\bar{D}$ can be cut open to produce a plane model.

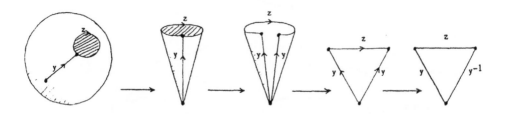

Fig. 11.1.4

So we see a word for $\bar{D}$ is

$$\bar{D} = yzy^{-1} ,$$

and this is the **normal form** for $\bar{D}$.

Combining this with Theorem 11.1.1 we can see that an orientable surface with boundary can have a word representation

$$a_1 b_1 a_1^{-1} b_1^{-1} \cdots a_n b_n a_n^{-1} b_n^{-1} y_1 z_1 y_1 \cdots y_m z_m y_m^{-1} \ ,$$

and a non-orientable surface with boundary can have a word representation

$$a_1 b_1 a_1^{-1} b_1^{-1} \cdots \cdots a_n b_n a_n^{-1} b_n^{-1} c_1 d_1 c_1 d_1^{-1} y_1 z_1 y_1^{-1} \cdots \cdots y_m z_m y_m^{-1} \ ,$$

$$\text{or} \quad a_1 b_1 a_1^{-1} b_1^{-1} \cdots \cdots a_n b_n a_n^{-1} b_n^{-1} cc y_1 z_1 y_1^{-1} \cdots \cdots y_m z_m y_m^{-1} \ .$$

So the associated plane models may now have an odd number of edges.

As in Section 3.5 we can use these ideas as a basis for a combinatorial definition of surfaces.

DEFINITION 11.1.4   An m-gon with labelled edges, where each label appears at most twice, possibly with the index^{-1} attached to some, is called a **non-orientable m-gon** if its word representation has the form $\cdots a \cdots a \cdots$ for at least one edge a, and is called an **orientable m-gon** otherwise. The labels which occur singly correspond to what we call **boundary edges**.

We have seen above that surfaces with boundaries, considered as space models, can be cut open so that the plane models produced are orientable or non-orientable m-gons. So, working backwards, we use these plane models as a basis for a working mathematical model of surfaces with boundary. We define *combinatorial* surfaces with boundary as follows:

DEFINITION 11.1.5   An orientable surface is an orientable m-gon. A **non-orientable surface** is a non-orientable m-gon. We call this m-gon a **surface with boundary** if it contains at least one boundary edge.

Our intuitive ideas of surfaces in space led us to the initial Classification Theorem (Theorem 11.1.1). Now that we have built a working, combinatorial model for these spatial surfaces, we should mimic Theorem 11.1.1 to show that any combinatorial surface is homeomorphic to a surface with the expected normal form. Now that we are working with combinatorial surfaces, by "homeomorphic" we mean "can be converted to $\cdots$", by using operations 1 to 4. This extension of Theorem 3.6.1 takes the following form:

THEOREM 11.1.2   (The Classification Theorem for Plane Models)

(a) An orientable surface is homeomorphic to $(nT)(k\bar{D})$, for some $n \geq 0, k \geq 0$.

(b) A non-orientable surface is homeomorphic to $(nT) K (k\bar{D})$, or to $(nT) P (k\bar{D})$. for some $n \geq 0, k \geq 0$.

PROOF In Chapter 3 we showed how operations 1 to 4 could be used to gather together the T's, K's and P's in a word, and the possible presence of boundary edges in a word does not affect this process. So we can assume we have converted the word M into one of the forms $(nT)\,A$, $(nT)\,KA$, or $(nT)\,PA$, where A is some string of edges containing nothing of the form $\cdots a \cdots a \cdots$ or $\cdots a \cdots b \cdots a^{-1} \cdots b^{-1} \cdots$. The string A contains the discs $\bar{D}$, each of which has a word of the form yzy^{-1}, so how do we gather these together using operations 1 to 4?

If for some edge x, A has the form $\cdots x \cdots x^{-1} \cdots$ $\left(\text{or} \cdots x^{-1} \cdots x \cdots \right)$, then choose such a pair with as few edges between as possible. If there are no edges between then we can use operation 2 to remove the sphere xx^{-1}. Otherwise the edges between x and x^{-1} must occur singly, so replace this whole string of single edges by the letter u. Then, if we say $M = LA$, we have $A = Bxux^{-1}C$ (say) and so

$$
\begin{aligned}
M &= LBxux^{-1}C \\
&= ux^{-1}CLBx \quad \text{(op.1)} \\
&= uy^{-1}BCLy \quad \text{(op.3)} \\
&= Lyuy^{-1}BC \quad \text{(op.1)}
\end{aligned}
$$

Continuing in this way A will eventually be replaced by

$$
y_1 u_1 y_1^{-1} \cdots y_t u_t y_t^{-1} E, \quad \text{say.}
$$

If the string of edges E is empty then we are finished.

If not E contains only single edges so it can be replaced by a single edge v.

To convert this to the form for a disc we first of all join a new pair $q^{-1}q$ to the end of the word (using operation 2), to get

$$
\begin{aligned}
M &= Ly_1 u_1 y_1^{-1} \cdots y_t u_t y_t^{-1} vq^{-1}q \\
&= qLy_1 u_1 y_1^{-1} \cdots y_t u_t y_t^{-1} vq^{-1}
\end{aligned}
$$

Now q can be passed through L and then through the discs $y_i u_i y_i^{-1}$ as follows:

First q meets the words for the tori in L, in which case

$$
\begin{aligned}
M &= qaba^{-1}b^{-1}Q && \text{(say)} \\
&= a^{-1}b^{-1}Qqab && \text{(op. 1)} \\
&= a_1^{-1}qb^{-1}Qa_1b && \text{(op. 3)} \\
&= a_1^{-1}qb_1^{-1}a_1Qb_1 && \text{(op. 3)} \\
&= b_1a_1^{-1}qb_1^{-1}a_1Q && \text{(op. 1)} \\
&= b_2qa_1^{-1}b_2^{-1}a_1Q && \text{(op. 3)} \\
&= b_2^{-1}a_1Qb_2qa_1^{-1} && \text{(op. 1)} \\
&= b_2^{-1}a_2qQb_2a_2^{-1} && \text{(op. 3)} \\
&= b_2a_2^{-1}b_2^{-1}a_2qQ && \text{(op. 1)}
\end{aligned}
$$

Continuing in this way q passes through all of the tori in L.

If M is non-orientable q now meets a Projective plane or a Klein bottle. It passes through the words for the Projective planes as follows:

$$
\begin{aligned}
M &= NqaaQ && \text{(say)} \\
&= Na_1q^{-1}a_1Q && \text{(op. 4)} \\
&= Na_2a_2qQ && \text{(op. 4)}
\end{aligned}
$$

and passes through the words for the Klein bottles as follows:

$$
\begin{aligned}
M &= Nqabab^{-1}Q && \text{(say)} \\
&= Nqa_1a_1b^{-1}b^{-1}Q && \text{(op. 4)} \\
&= Na_3a_3b_2^{-1}b_2^{-1}qQ && \text{(passing through two} \\
& && \text{projective planes as above)} \\
&= Na_4b_2^{-1}a_4b_2^{-1}qQ && \text{(op. 4)}
\end{aligned}
$$

Finally q meets the words for the discs, which it can pass through as follows:

$$
\begin{aligned}
M &= Nqyzy^{-1}Q && \text{(say)} \\
&= y^{-1}QNqyz && \text{(op. 1)} \\
&= y_1^{-1}qQNy_1z && \text{(op, 3)} \\
&= Ny_1zy_1^{-1}qQ && \text{(op. 1)}
\end{aligned}
$$

So finally q has passed through almost the whole word M, giving

$$
M = Ly_1u_1y_1^{-1}\cdots y_tu_ty_t^{-1}qvq^{-1} ,
$$

and the proof of the theorem is complete.

11.2   RECOGNISING SURFACES WITH BOUNDARY

Theorem 11.1.2 shows that the number of discs in the normal form, which is the number of cuffs in the surface, is a vital piece of information needed to classify the surface. In the *normal form* this number is the number of edges which occur exactly once, i.e. it is the number of boundary edges.

However any edge x in a plane model could be divided into a sequence of edges $x_1 x_2 \cdots x_m$ provided the second edge $(x$ or $x^{-1})$, if it exists, is similarly divided. Such changes have no effect on the space model produced when the edges are identified in pairs. So, as can be seen in Fig. 11.2.1 the boundary of a cuff in a surface may be represented by a single boundary edge x, as in Fig. 11.2.1(a), or by a sequence of boundary edges $x_1, x_2 \cdots x_m$ as in Fig. 11.2.1(b).

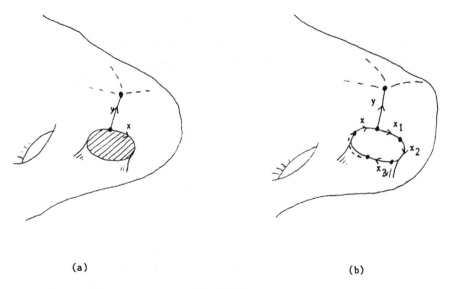

(a)   (b)

Fig. 11.2.1

It follows from this that the number of cuffs is not necessarily equal to the number of boundary edges. So given a plane model, or its associated word, how do we discover the number of cuffs the surface contains?

The first thing we must do is determine the distinct vertices in the plane model, using the technique of Section 4.1. Next we list the boundary edges and their associated vertices. We can now see how these boundary edges fit together to form the boundaries of cuffs, which we shall call **boundary curves**. This process is illustrated in the following example.

EXAMPLE 11.2.1   Consider $M = abcd^{-1}a^{-1}ec^{-1}$. This surface has boundary edges b, d^{-1} and e.

Label the vertices in the plane model, as shown in Fig. 11.2.2

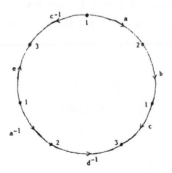

Fig. 11.2.2

Now list the boundary edges, with their vertices, as in Fig. 11.2.3(a). It can be seen that these boundary edges fit together to form a single boundary curve, as in Fig. 11.2.3(b). This boundary curve is the boundary of the single cuff in the surface M..

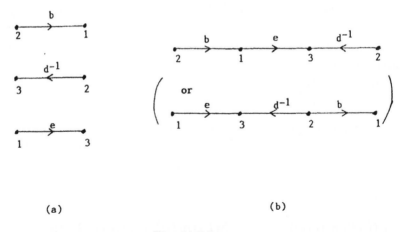

(a)   (b)

Fig. 11.2.3

In general, since it is clear that Operations 1 to 4 do not affect the number of boundary *curves* in a word, and since the number of boundary curves in the *normal form* is equal to the number of cuffs, we get the following theorem. THEOREM 11.2.1   The number of boundary curves in a surface is the same as the number of cuffs in the normal form of the surface.

EXAMPLE 11.2.1 illustrated how to determine the number of boundary curves in a given word, and therefore the number of cuffs in the normal form for the surface. In this particular case it is now easy to recognise the normal form. The torus $aca^{-1}c^{-1}$ can be spotted in M, and now all edges have been accounted for, so the normal form for M must be $T \# \bar{D}$.

In general we could always discover the normal form hidden in the rest of the word using algebra and operations 1 to 4. However for complicated surfaces this can be tedious. A much better way is to use the Euler characteristic as in Chapter 4, so let's extend the relevant results to do this. First the extension to Corollary 4.2.1.

LEMMA 11.2.1

$$
\begin{array}{lll}
\text{(a)} & \chi\left((nT)\left(m\bar{D}\right)\right) & = & 2 - 2n - m \\
\text{(b)} & \chi\left((nT)\,K\left(m\bar{D}\right)\right) & = & -2n - m \\
\text{(c)} & \chi\left((nT)\,P\left(m\bar{D}\right)\right) & = & 1 - 2n - m
\end{array}
$$

PROOF   First we need $\chi\left(\bar{D}\right)$.

$\bar{D} = yzy^{-1}$ so, as an exercise, sketch the plane model for $\bar{D}$, find the number of distinct vertices for this model, and then see that $\chi\left(\bar{D}\right) = v - e + 1 = 1$.

We can show $\chi\left(m\bar{D}\right) = 2 - m$ by induction, as follows.

The case $m = 1$ has been dealt with in the first part of the proof. Suppose the result is true for $m = k$, and consider $(k+1)\,\bar{D}$.

Then $\chi\left((k+1)\,\bar{D}\right) = \chi\left((k\bar{D})\,\#\bar{D}\right) = (2 - k) + 1 - 2$ (by Theorem 4.2.1.)

An appeal to induction therefore proves that $\chi\left(m\bar{D}\right) = 2 - m$ for each positive integer m.

Now, to prove:

(a)

$$
\begin{array}{lll}
\chi\left((nT)\left(m\bar{D}\right)\right) & = & \chi\left(nT\right) + \chi\left(m\bar{D}\right) - 2 \quad \text{(by Theorem 4.2.1)} \\
& = & (2 - 2n) + (2 - m) - 2 \quad \text{(by Corollary 4.2.1)} \\
& = & 2 - 2n - m,
\end{array}
$$

and (b)

$$
\begin{array}{lll}
\chi\left((nT)\,K\left(m\bar{D}\right)\right) & = & \chi\left((nT)\left(m\bar{D}\right)\right) + \chi(K) - 2 \quad \text{(by Theorem 4.2.1)} \\
& = & 2 - 2n - m + 0 - 2 \quad \text{(using (a) above)} \\
& = & -2n - m,
\end{array}
$$

and (c)

$$
\begin{array}{lll}
\chi\left((nT)\,P\left(m\bar{D}\right)\right) & = & \chi\left((nT)\left(m\bar{D}\right)\right) + \chi(P) - 2 \\
& = & 2 - 2n - m + 1 - 2 \\
& = & 1 - 2n - m.
\end{array}
$$

We can now extend Theorem 4.3.1 as follows:

THEOREM 11.2.2   The surfaces M_1 and M_2 are homeomorphic if and only if

(a) $\chi\left(M_1\right) = \chi\left(M_2\right)$,

and (b) M_1 and M_2 have the same orientation,

and (c) M_1 and M_2 have the same number of boundary curves.

PROOF   As in the proof of Theorem 4.3.1 we see that operations 1 to 4 do not affect (a), (b) or (c), so if M_1 and M_2 are homeomorphic then they do satisfy (a), (b) and (c).

Conversely, if the surfaces M_1 and M_2 are orientable and each have m boundary curves then, by Theorems 11.1.2 and 11.2.1, M is homeomorphic to $(n_1 T)(m\bar{D})$ and M_2 is homeomorphic to $(n_2 T)(m\bar{D})$, for some integers n_1 and n_2. Then by lemma 11.2.1

$$\chi(M_1) \;=\; 2 - 2n_1 - m$$
$$\text{and} \quad \chi(M_2) \;=\; 2 - 2n_2 - m.$$

So if $\chi(M_1) = \chi(M_2)$ we see $n_1 = n_2$ and so M_1, M_2 are each homeomorphic to $(n_1 T)(m\bar{D})$. So M_1 and M_2 are homeomorphic.

Similarly if M_1, M_2 are non-orientable and each have m boundary curves then,

$$M_1 \;=\; (n_1 T) K (m\bar{D}) \quad \text{or} \quad M_1 \;=\; (n_1 T) P (m\bar{D}),$$
$$\text{and} \quad M_2 \;=\; (n_2 T) K (m\bar{D}) \quad \text{or} \quad M_2 \;=\; (n_2 T) P (m\bar{D}).$$

If $\chi(M_1) = \chi(M_2)$ and $\chi(M_1) + m$ is even, we must have $\chi(M_1) = -2n_1 - m$ and $\chi(M_2) = -2n_2 - m$, in which case $n_1 = n_2$ and $M_1 = M_2 = (n_1 T) K (m\bar{D})$.

If $\chi(M_1) = \chi(M_2)$ and $\chi(M_1) + m$ is odd, we must have $\chi(M_1) = 1 - 2n_1 - m$ and $\chi(M_2) = 1 - 2n_2 - m$, and so $n_1 = n_2$ and $M_1 = M_2 = (n_1 T) P (m\bar{D})$.

This completes the proof.

Now we are in a position to quickly and easily determine the normal form of any surface, with or without boundary, given as a plane model or word.

To do this:

- check the orientability of the surface

- sketch the plane model and determine the number of distinct vertices

- evaluate the Euler Characteristic

- determine the number of boundary curves

- finally match the Euler characteristic to the relevant formula in Lemma 11.2.1 and read off the normal form for the surface.

EXAMPLE 11.2.2   Find the normal form for

$$M = abcdea^{-1}fgd^{-1}he^{-1}f^{-1}c^{-1} \ .$$

Solution   The plane model, with distinct vertices labelled, is:

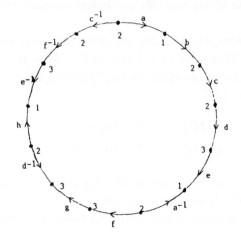

So $v = 3, e = 8$ and $\chi(M) = 3 - 8 + 1 = -4$.
The boundary edges, and boundary curves are:

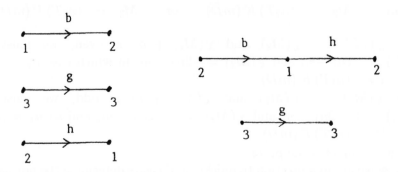

Boundary Edges   Boundary Curves

So the number of boundary curves is 2.
Now M is an orientable surface so the formula

$$\chi(M) = 2 - 2n - m = -4 \ , \ \text{where } m = 2 \ , \text{ gives } n = 2 \ .$$

So the normal form for M is $(2T)(2\bar{D})$.

11.3   AN APPLICATION TO KNOTS

We think of a knot as an "infinitesimal thin thread" interlaced with itself,
and with the loose ends joined together. An example is shown in Fig. 11.3.1.
Notice how this knot, which we imagine to exist in three-dimensional space,
has been projected onto two-dimensional space when we have drawn it on
the page.

Notice also that in sketching this two-dimensional projection an obvious convention has been used to explain which part of the thread is on top at any apparent crossing.

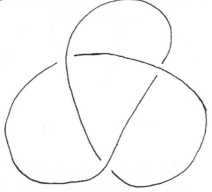

Fig 11.3.1   A Trefoil Knot

The Theory of Knots is a fascinating subject. The decorative appeal of knots has been evident in many art forms through the ages, and there are also applications to be found in Biology and Chemistry. Recent developments have also uncovered connections with Quantum Field Theory, and Knot Theory has once again become an active research area. For anyone interested in learning more about knots the books listed at the end of this chapter are warmly recommended.

Here our aim is to give just a glimpse of a link between knots and surfaces. As with the Theory of Surfaces, the big problem in Knot Theory is that of classification. Can we establish that two given knots are different? Or can we show that two given drawings actually represent the same knot, by matching them to some list of "normal forms"?

Now the strands which form the knots consist of nothing more than a closed loop in three-dimensional space, so they are all homeomorphic to a circle. The true nature of the knots is determined by the way the strands are interlaced in space, so it is determined by the space between and around the strands. One way to include the space between the strands is to construct a surface with the knot as a single boundary curve. To see how this is possible consider the trefoil in Fig. 11.3.1.

First we give the knot an orientation by choosing a direction to travel around the strand. For the trefoil this is shown in Fig. 11.3.2.

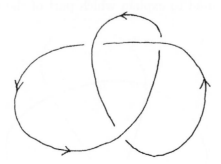

Fig. 11.3.2   An oriented trefoil

Next at each crossing we create a "short circuit", following the arrow directions, as shown in Fig. 11.3.3.

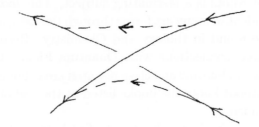

Fig. 11.3.3   Bypassing a crossing

In this way each crossing is removed.

We do this for each of the three crossings in Fig. 11.3.2 and, get the two circuits shown in Fig. 11.3.4. These are called the **Seifert circles**.

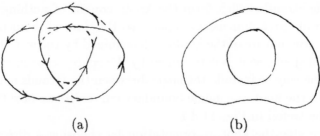

(a)   (b)

Fig. 11.3.4   The Seifert circles for the trefoil

The Seifert circles are boundaries of discs, and these discs are now joined to construct the required surface. At each of the original crossings we make a join as shown in Fig.11.3.5.

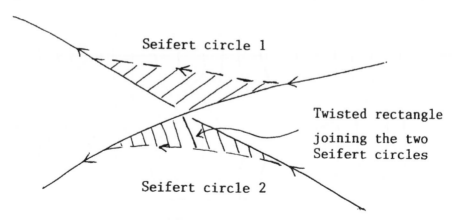

Fig. 11.3.5

This join is made by means of a twisted rectangle as described by the crossing. By considering the orientations carefully in this way the surface produced is *orientable*, and its boundary will be the original trefoil. Seiferts
Theorem states that this procedure can be carried out for any knot, and
moreover that from information gleaned from the drawing of the knot we
can determine the normal form for a surface which has the knot as its single
boundary curve.

THEOREM 11.3.1 (Seifert)   Given any knot K in $\mathbb{R}^3$ there is a space model
for an orientable surface M whose single boundary curve is the knot K.
If the two-dimensional drawing of K has x crossings, and the procedure
above produces c Seifert circles, then the surface produced has normal form
$\left(\dfrac{1}{2}\left(x - c + 1\right)T\right)\bar{D}.$

EXAMPLE 11.3.1   For the trefoil above there were 3 crossings and we
produced 2 Seifert circles, so the orientable surface constructed having the
trefoil as its boundary has normal form $\dfrac{1}{2}\left(3 - 2 + 1\right)T\bar{D} = T\bar{D}$ as its normal
form.

11.4 EXERCISES

1.   Fig 11.4.1 shows space models of three surfaces with boundaries. (Think
of them as being constructed of an infinitesimally thin transparent film.) By
colouring the boundaries determine the number of boundary curves for each
surface. Explain why no two of these surfaces could be homeomorphic.

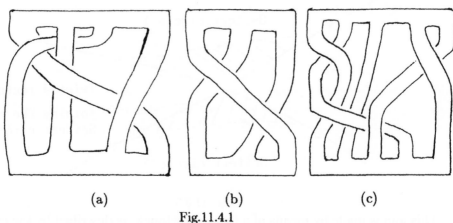

(a)   (b)   (c)

Fig.11.4.1

2. Find the Normal Form for each of the following surfaces with boundary. Do this by applying operations 1 to 4, and then check your answers by determining the number of boundary curves and the Euler characteristic and then looking for the appropriate match in Lemma 11.2.1.

(a) The Möbius band, $abac$.

(b) $abcda^{-1}d^{-1}c^{-1}eb^{-1}$.

(c) $abcdad^{-1}ceb^{-1}$.

(d) $abcb^{-1}dcefa^{-1}e^{-1}$.

(e) $abcadbcefe$.

3. Determine the number of Seifert circles for each of the knots shown in Fig. 11.4.2. In each case use Theorem 11.3.1 to find the normal form of a surface with this knot as its boundary.

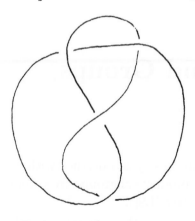

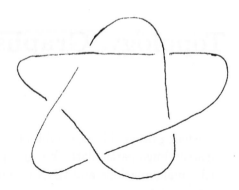

a) A figure-eight knot   (b) A cinquefoil knot

Fig.11.4.2

Further Reading

Adams, C.C. (1994) *The Knot Book*, W.H.Freeman and Co.

Gilbert, N.D. & Porter T. (1994) *Knots and Surfaces*, Oxford University Press

12

Topology, Graphs and Groups

Among other things, combinatorial group theory is concerned with group constructions, particularly free groups, free products of groups, free products with amalgamations and HNN extensions. (see [LS]).

It is of interest to know how the subgroups of these groups are related to the original constructions. For example, subgroups of free groups are themselves free groups, and subgroups of free products are themselves free products. The first proofs of these results due to Nielson, Schreier and, for free products, to Kurosh. (see [K] volume II) were obtained by purely algebraic methods involving manipulation of words and consideration of how words and parts of words cancel in products. These methods, although showing in detail, what is going on in the group, are very tedious. We have seen how the fundamental group can give information about a topological space. The converse also happens. It is the exploitation of this idea which has led to powerful methods of exploring the problems mentioned above. For details see [L], [M], [S] and [ZVC]. It is not feasible to pursue this approach much further here but we can say a little more about the Nielsen-Schreier theorem on free groups.

Nielson-Schreier Theorem   If G is a free group with k free generators and H is a subgroup of index n in G, then H is free on $(k-1)n+1$ free generators. (We discuss the idea of index later).

The tessellations considered in Chapter 8, in particular Example $8 \cdot 4 \cdot 2$ for K the Klein bottle, provide covering spaces for the compact surfaces considered.

This is a particular example of the general concept of the covering space of a given space. (see [L]). The vital fact, as far as we are concerned, is that the fundamental group of the covering space is a subgroup of the fundamental group of the base space.

In the case of free groups these spaces are particularly simple. They are one-dimensional i.e. graphs, and the fundamental groups are all free. From this approach the Nielsen-Schreier theorem is fairly easy to deduce. By using more complicated spaces the Kurosh theorem for the structure of the subgroups of free products may be deduced in this way. (see [M]).

Note that the free product G of groups A and B is

$$G = <a_1, \cdots, a_n, \ b_1, \cdots, b_m \mid R_1, \cdots, R_t, \ S_1, \cdots, S_\ell >, \text{ where}$$
$$A = <a_1, \cdots, a_n \mid R_1, \cdots, R_t > \text{ and}$$
$$B = <b_1, \cdots, b_m \mid S_1, \cdots, S_\ell > \ .$$

In other words, G is generated by A and B with **no** further relations between the generators other than those already existing in A and B separately.

Purely algebraic methods have been obtained from this topological approach by abstracting the algebra involved. These various algebraic methods are discussed in [C], [H], and [R]. In [C] Cohen also compares some of the methods. The ultimate method obtained along these lines appears to be the Bass-Serre Theory of groups acting on graphs, particularly trees. This method has led to an understanding of the structure of many groups by allowing the group G to act on a tree and then from this constructing a graph of groups whose fundamental group is G. From this it is possible to deduce the structure of subgroups of G, and also in certain cases to discover how G may be constructed from some of its subgroups. (see [B], [C], [D], [S]).

In stating the Nielsen-Schreier theorem we mentioned the **index** of a subgroup H in the group G.

Let us consider this and the idea of a permutation representation of a group G relative to a subgroup H. This will be needed later.

Suppose that H is a subgroup of G. Define a **left coset** of H in G to be the set: $\{gH \mid g \in G\}$.

If H is finite, say, $H = \{h_1, h_2, \cdots, h_m\}$, then $gH = \{gh_1, gh_2, \cdots, gh_m\}$. Similar results hold for a **right coset** of H in G. We have $\{Hg \mid g \in g\}$.

The left (right) cosets of H in G, gH; as g runs through the elements of G, partition G.

This means that $g_1 H \cap g_2 H = \phi$ or $g_1 H = g_2 H$. In other words if two cosets have any elements in common then they are identical.

To show this, suppose that $x \in g_1 H \cap g_2 H$ for $g_1 = g_2$. Then $x = g_1 h_1 = g_2 h_2$ for some $h_1, h_2 \in H$.

Now let y be any element of $g_1 H$, say $y = g_1 h, h \in H$. We can write $y = \left(x h_1^{-1}\right) h = \left(g_2 h_2 h_1^{-1}\right) h \in g_2 H$. Thus $g_1 H \subseteq g_2 H$; that is $g_1 H$ is a subset of $g_2 H$. Similarly $g_2 H \subseteq g_1 H$. Hence $g_1 H = g_2 H$.

An exactly parallel proof shows that the right cosets Hg also partition G.

However, in general, the two partitions are different. When the two partitions are identical so that left and right cosets coincide, the subgroup H is a very important special subgroup called a **normal** subgroup. When H is normal the set of cosets can be given the structure of a group by defining the product of two cosets by: $(g_1 H) \circ (g_2 H) = (g_1 g_2 H)$. The resulting group is called the quotient or factor group and written $\dfrac{G}{H}$. (see [G1]).

If G is finite and the **distinct** left cosets are $g_1 H, g_2 H, \cdots, g_r H$, then, because $|g_i H| = |H|$ for $i = 1, 2, \cdots r$, where $|S|$ denotes the number of elements in S, it follows that $r|H| = |G|$. Thus the order of the subgroup H divides the order of the group G. This is known as Lagrange's Theorem discovered in the very early days of group theory. From this we can deduce quite easily that the order of each element of G divides the order of G, where order of G is defined to be $|G|$. The order t of an element g is the smallest t so that $g^t = e$ (the identity of the group).

The number r in the above (i.e. the number of cosets of H in G) is called the index of H in G. This is usually written [G : H] and is the same for both left and right cosets.

Closely associated with the idea of the coset is a permutation representation of a group G, in particular by permutations of the cosets relative to a subgroup H. We have the following.

Let f be a function from the group G into the group H. If $f(g_1 g_2) = f(g_1) f(g_2)$, for any $g_1, g_2 \in G$, then f is called a homomorphism from G into H. This is a generalisation of the idea of an isomorphism mentioned earlier in this book. In fact if $f(g_1) = f(g_2)$ implies that $g_1 = g_2$ **and** if for **any** $h \in H$ we have $f(g) = h$ for **some** $g \in G$, then f is an isomorphism of G with H. However we can always get an isomorphism from a homomorphism as follows.

Let $Imf = \{h \in H | f(g) = h$ for some $g \in G\}$ denote the image of f. That is the set of all elements of H which are mapped onto by the function f. Also let $Kerf = \{g \in G | f(g) = e\}$, where e is the identity of H, denote the kernel of f, which is the set of all the elements of G that are mapped by the function f onto the identity of H. Then it turns out that the factor group (see earlier) $\dfrac{G}{Kerf}$ is isomorphic to Imf. We write $\dfrac{G}{Kerf} \cong Imf$. (see [G1]).

Any homomorphism f of a group G into the symmetric group S_n (i.e. the group of all permutations on n symbols) is called a permutation representation of G. A particularly important permutation representation is obtained as follows.

Let H be an arbitrary subgroup of G. Let S_n be the symmetric group of all permutations on the n left cosets of H relative to G, where the index of H in G, $[G : H] = n$. Define a function $f : G \to S_n$ by $f(g) = P_g$, where P_g is the permutation on the n left cosets $x_1 H, x_2 H, \cdots, x_n H$; $x_i \in G$; given by

$$
P_g = \begin{pmatrix} x_1 H & x_2 H \cdots & x_n H \\ g x_1 H & g x_2 H \cdots g x_n H \end{pmatrix} .
$$

i.e. $P_g(x_i H) = g x_i H$.

Now

$$
f(g_1 g_2) = P_{g_1 g_2} = \begin{pmatrix} x_1 H & \cdots & x_n H \\ g_1 g_2 x_1 H & \cdots & g_1 g_2 x_n H \end{pmatrix}
$$

$$
= \begin{pmatrix} g_2 x_1 H & \cdots & g_2 x_n H \\ g_1 g_2 x_1 H & \cdots & g_1 g_2 x_n H \end{pmatrix} \begin{pmatrix} x_1 H & \cdots & x_n H \\ g_2 x_1 H & \cdots & g_2 x_n H \end{pmatrix}
$$

$$
= P_{g_1} P_{g_2} = f(g_1) f(g_2) .
$$

Hence f is a homomorphism and we have obtained a permutation representation of G.

We can say something about the kernel of f in the above. If N is any normal subgroup of G such that $N \subseteq H$, then $N \subseteq Ker f \subseteq H$.

This means that $Ker f$ is the 'largest' normal subgroup of G contained in H. We can see this as follows.

$g \in Ker f$ if and only if $P_g =$ the identity permutation, which is so if and only if $g x_i H = x_i H$ for all i, and this holds if and only if $g x H = x H$ for all $x \in G$.

Taking $x = e$, we have $gH = H$. Hence $g \in H$. This means that $Ker f \subseteq H$. Now take N to be any normal subgroup of G such that $N \subseteq H$. Let $a \in N$ and $g \in G$. Then $g^{-1} a g \in N \subseteq H$, because N is normal. In particular $g^{-1} a g \in H$. Hence $g^{-1} a g H = H$. Thus $a g H = g H$. But g is **any** element of G. Thus $a \in Ker f$, and we conclude that $N \subseteq Ker f$. Altogether we have $N \subseteq Ker f \subseteq H$, as required.

A well-known special case of the above is usually referred to as Cayley's Theorem which may be stated as:

Any finite group G is isomorphic to a subgroup of the symmetric group S_n for some $n \leq |G|$. Just take $H = \{e\}$ in the previous work. Then we have a homomorphism $f : G \to S_n$, where $n = |G|$ and $Ker f \subseteq H = \{e\}$. Hence $Ker f = \{e\}$.

But $Kerf = \{e\}$ implies that f is one-one i.e. $f(g_1) = f(g_2)$ implies $g_1 = g_2$, and so f is an isomorphism of G onto some subgroup of S_n.

Alternatively, we could quote the result mentioned earlier and conclude that

$$Imf \cong \frac{G}{Kerf} = \frac{G}{\{e\}} \cong G .$$

When the group G is given in terms of generators and relations, as discussed in Chapter 10, we can obtain a permutation representation of G relative to the cosets of a subgroup H of G by a systematic enumeration of cosets using an algorithmic procedure due to Todd and Coxeter and named the Todd-Coxeter algorithm in their honour.

This can be implemented on a computer, and can be made to yeild a set of generators and relations to describe the subgroup H. This is discussed in detail in [CM] and in [G2].

This enumeration can be illustrated graphically by a Schreier diagram, which is a generalisation of a Cayley graph discussed earlier in Chapter 8. Instead of the vertices of the graph representing elements of the group, they now denote the cosets of H in G. Otherwise the graph is produced in the same way. An example due to Dehn (a pioneer in this branch of algebra) is constructed in [Bo] for the group

$$G = < a, b | a^2 = b^2 = (ba)^3 = e >$$

relative to the subgroup $H = < b | b^2 = e >$. This gives the diagram below.

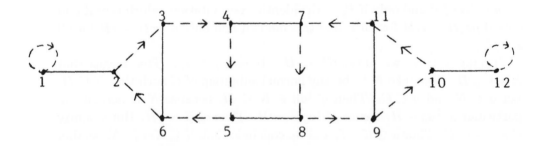

where ——— denotes the generator a
and – – – denotes b. The numbers $1 \longrightarrow 12$ label the cosets.

As with coset enumeration if we want the Schreier diagram to carry more information then we fix special representatives for the cosets, and label the diagram with these, and then keep track of the effect of the generators on these representatives. By selecting a spanning tree (i.e. a subgraph including all vertices and such that any pair of vertices is connected by a unique path) of the Schreier diagram as a basis for the choice of coset representatives we can obtain a special set of representatives called a Schreier system. Then from a Schreier diagram we can read off a set of generators for H and a set of defining relations which can be written in terms of these generators by a procedure called a Reidemeister-Schreier rewriting process. For a nice concise description of this see [Bo].

The above methods of finding generators and relations for a subgroup H of a group G can be performed without reference to diagrams by a purely algebraic process. This is described in detail in [MKS] and [J]. However, particularly for subgroups of small index $[G : H]$ in G, the graphical and coset enumeration techniques provide especially illuminating and simple approaches.

References for Chapter 12

[A] Baumslag,G (1993) *Topics in Combinatorial Group Theory*, Birkhäuser.

[Bo] Bollobás,B (1979) *Graph Theory*, Springer-Verlag.

[C] Cohen,D.E. (1989) *Combinatorial Group Theory: a topological approach*, Cambridge University Press.

[CM] Coxeter,H.S.M. and Moser,W.O.J. (1980) *Generators and Relations for Discrete Groups* (Fourth Edition), Springer-Verlag.

[D] Dicks,W and Dunwoody,M.J. (1989) *Groups Acting on Graphs*, Cambridge University Press.

[G1] Gardiner,C.F. (1980) *A First Course in Group Theory*, Springer-Verlag.

[G2] Gardiner,C.F. (1986) *Algebraic Structures*, Ellis Horwood Ltd.

[H] Higgins, P.J. (1971) *Categories and Groupoids*, Van Nostrand Reinhold Co.

[J] Johnson,D.L. (1990) *Presentations of Groups*, Cambridge University Press.

[K] Kurosh,A.G. (1955) *The Theory of Groups Vol. I*, (1956) Vol II, Chelsea Publishing Co.

[L] Lee,J.M. (2000) *Introduction to Topological Manifolds*, Springer-Verlag.

[LS] Lyndon,R.C. and Schupp,P.E. (1977) *Combinatorial Group Theory*, Springer-Verlag.

[MKS] Magnus,W. Karrass,A and Solitar,D. (1966) *Combinatorial Group Theory*, Wiley - Interscience.

[M] Massey,W.S. (1967) *Algebraic Topology: An Introduction*, Harcourt, Brace and World, Inc.

[R] Rotman,J.J. (1995) *An Introduction to the Theory of Groups* (Fourth Edition), Springer-Verlag.

[S] Serre,J-P. (1980) *Trees*, Springer-Verlag.

[St.] Stillwell,J. (1980) *Classical Topology and Combinatorial Group Theory*, Springer-Verlag.

[ZVC] Zieschang,H. Vogt,E and Coldewey,H-D. (1980) *Surfaces and Planar Discontinuous Groups*, Springer-Verlag.

Outline solutions to the exercises

EXERCISES 1.11

(1) (a) The sphere. This is a compact surface.
   (b) A sphere with a hole, including the 1-dimensional boundary. The set is connected and bounded, but it is not a 2-dimensional manifold.
   (c) A sphere with a hole, excluding the 1-dimensional boundary. Connected, bounded and a 2-dimensional manifold, but not a compact surface.
   (d) Two quadrants of the plane, excluding 1-dimensional boundaries. A 2-dimensional manifold, but not connected and not bounded.
   (e) The torus. This is a compact surface.
   (f) A solid octahedron. Connected, bounded, but not a 2-dimensional manifold.

(2) (a) No. (b) Yes.

(3) See the solution to Exercise 2.7.2, and glue together the sides of the plane model.
   No.

(5) (b) A, D, O, P, Q, R; (c) B; (a) the rest.

(9) ├────────────┤ and ∞ are not 1-dimensional manifolds.
   ├────────────→ is a 1-dimensional manifold, but is not 'bounded and without zero-dimensional boundaries'

   is a bounded, 1-dimensional surface with no zero-dimensional boundaries.

(10) Read on to chapter 4.

EXERCISES 2.7

(2)

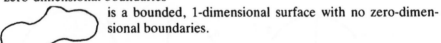

(3) (a) gcd(n,p). (b) Only when q is rational. (c) Always. (d) (i) rational r (ii) r an even integer.

(4) (a) Read on and compare with Fig. 3.1.2.

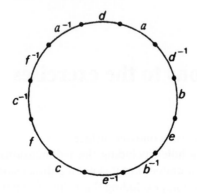

   (b) See Fig. 3.1.2.

(5) Read on to section 4.1.

EXERCISES 3.8

(3) (a) $a_1a_2a_1^{-1}a_2^{-1}b_1b_2b_1^{-1}b_2^{-1}c_1c_2c_1c_2^{-1}d_1d_2d_1d_2^{-1}$

   (b) $a_1a_2a_1^{-1}a_2^{-1}b_1b_2b_1^{-1}b_2^{-1}c_1c_2c_1^{-1}c_2^{-1}ddee$

   (c) $a_1a_2a_1a_2^{-1}b_1b_2b_1b_2^{-1}c_1c_2c_1c_2^{-1}ddee$

(4) For any basic surface, except S, the vertices correspond to a single point. In the connected sum construction, for any pair, these points are identified; see Fig. 3.1.1.

(5) (b)

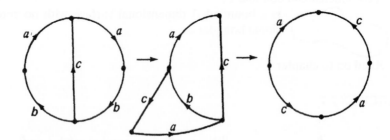

(6) (a) See the proof of Theorem 3.6.1.

   (b) $3TK$.

(7) Use induction based on Theorem 3.6.1(b) and exercise (6).

(8) (a) $2T$; (b) not a 2n-gon; (c) K; (d) TP;

   (e) Möbius band (not a compact surface);

   (f) $3TP$.

(10) Four, and anyone who can't see it is a doughnut!

EXERCISES 4.6

(2) (a) $3TK$; (b) $3T$; (c) S; (d) $6TK$.
(4) (a) Use Theorem 4.2.1.
   (b) $M \equiv 4P3KT \equiv 5TK$.
(6) Removing handles, a layer at a time gives
   (a) $n^2(2n + 3)$; (b) $3n^2 + 9n + 5$.
(7) (a) $(n/2)(n - 1)$.
   (b) $n2^{n-1} - 2^n + 1$
(8) (a) (i) $(2n - 1)!$ (ignoring rotations).
       (ii) $(2^n - 1)(2n - 1)!$ (ignoring rotations).
   (b) (i) $[n/2] + 1$ (For notation see section 6.1).
       (ii) n.
(10) (a) $KKSS \equiv PPKS \equiv PPPP \equiv PTPS \equiv TKSS$
   (b) $2n + 1$.
   (c) (i) $6TK$. (ii) $15TP$. (iii) $7TP$.

EXERCISES 5.6

(1) Read on. Chapter 9 provides a useful technique for checking patterns.
   (a) $P_\sqrt{}\ C\times$. (b) $C_\sqrt{}$. (c) $P\times$. (d) $C_\sqrt{}$. (e) $C_\sqrt{}$.
(2) Octahedron.
(4) Fig. 5.6.1(b) shows the complex of a tetrahedron drawn on a plane model of S.
(8) (a) Superimpose a pattern of smaller hexagons as follows:

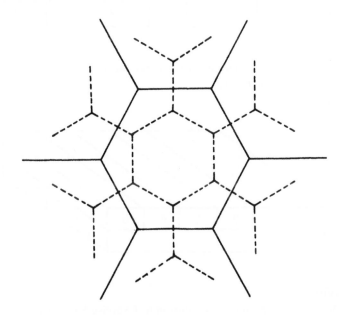

(b) Sketch the duals of the examples of $(6, 3)T$.

(c) Divide the rectangles of Fig. 5.6.5(a) into triangles in a suitable fashion.

(9) (a) From equation (1) of section 5.2.

(11) 999.

(12) (a) Eliminate F_6 from equations (3) and (4) of section 5.2.

Yes. See the Platonic solids.

(b) For a space model, shave one corner and three edges from a cube.

(c) For a space model, place a quadrilateral at each vertex of Fig. 5.6.5.

(13) For an example of such a complex, divide one of the pentagons in $(5,3)P$ into triangles.

(14) Equation 7 of section 5.4 leads to $F_3(6 - b) + F_4(8 - 2b) + F_5(10 - 3b) = 0$ and consideration of the various possibilities for b gives the result.

(15) Let $2F$ be the total number of faces on any such complex. The equations in the text lead to $6 = F(12 - (n + m))$. From this we see that there are just two possible values for F. For each of these values there are three distinct possibilities for the unordered pair (m,n). But not all of these give realizable complexes.

EXERCISES 6.9

(1)

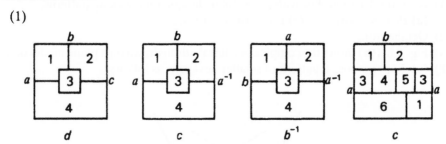

(2) A central face surrounded by a ring of five faces.

(3) False. Extend Table 6.1.1.

(4) Consider.

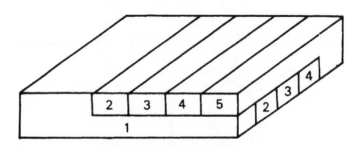

(6) $a \leq n[b/n]$.

(7) (a) Dodecahedron. See your solution to Exercise 5.6.4(b).

(b) Look for subgraphs homeomorphic to $K_{3,3}$.

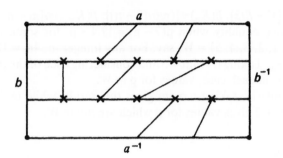

(8) (a) See Exercise 4.6.7(a).
  (b) $(m-1)(n-1)$.
(9) Place m vertices around the equator, and the other two at the poles.
(10) (b) Suppose, on the contrary, that for some n we can embed K_{4+7n} in $4n^2T$. Then $4+7n$ colours are needed to colour the vertices of K_{4+7n} (or, if you like, the map dual to the pattern formed by the embedding), so

$$4+7n \leqslant [(7+\sqrt{49-24\chi})/2]$$

and this leads to the result.
(11) (a) For K_7 see your solution to Exercise 5.6.8(b).
  (b) Dualize the appropriate patterns produced in (a).
(12) (b) $V=2^n$, $E=n2^{n-1}$, and each face has at least four edges. Thus

$$\gamma(C_n) \leqslant 2^{n-2}(4-n).$$

  (c) A suitable embedding shows $\gamma(C_4)=0$.
(13) $\gamma(G) \leqslant 2[((1+m+n)-(1m+mn+ln)/3)/2]$

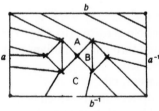

(14)

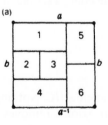

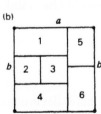

(a)   (b)

and the dual   and the dual

(15) (a) $\beta(G \leqslant [V - E/3]$; $\beta(K_n) \leqslant [(n/6)(7 - n)]$; $\beta(K_{m,n}) \leqslant [m + n - mn/2]$.

(d) We have equality when $n(7 - n) = 12N + p$, for some integer N, where $p \in \{0, 1, 2, 3, 4, 5\} = W$, say. For any integer m, $4m = 12k + q$, where $q \in \{0, 4, 8\}$. Taking $n = 4m + 3$ and considering each of the possible values for q gives, in each case, values for p in W.

Similarly, taking $n = 4m$ gives this result, while taking $n = 4m + 1$ or $n = 4m + 2$ gives values for p which are *not* in W.

EXERCISES 7.9

(1) Base the contours on the perpendicular height from a 'smoothed out' form of the planet.

(4) -2.

(5) By definition, the index is an integer. The result then follows from the formula $1 + (a - c)/2$.

(6) Possible cases are:

	a	b	c
	4	0,1,2	0
	4	0	1

(7) The rotation of the earth applies a twisting force to water falling down a sink. The same rotation caused the movement of shadows on sundials, in the northern hemisphere, on which clock faces are based.

(9) (a) No critical points.

(b) One critical point, index 1, at the origin. A centre.

(c) Saddle points at $(\pm 1, 0)$.

(d) Saddle point.

(e) A centre and a saddle point.

(f) A sink at $(0, 0)$ and saddles at $(1, \pm 1)$.

(g) Four centres and a central saddle.

(h) A single critical point with six hyperbolic sectors.

(10) (a) $x = Ky^3$ (for varying constants K)

(b) $y^a e^{-by} = Kx^{-c}e^{dx}$.

(11) (a) $y - 2x$; (b) $-(x^2 + y^2)/2$; (c) $y - x^2y$; (d) $x(y - (x^2/3))$; (e) $(y^2/2) + x((x^2/3) - 1)$; (f) and (g) No ϕ exists; (h) $x^2y - xy^3$.

(12) (a) $V^*(p)$ is $V(p)$ rotated through $\pi/2$ radians anti-clockwise.

(c) The arrows are reversed.

(d) Lines of steepest ascent, etc.

(13) In Chapter 9, we develop techniques which help solve this type of problem.

(a) A source in the middle, a sink on one edge, a saddle on the other edge and at the vertex.

(b) Two centres.

EXERCISES 8.6

(1) We have assumed that D lies inside the triangle ABC. Nothing in Euclid's axioms stops us from doing this. A complete set of axioms removes all dependence on diagrams.

(2) (b) Translation along the first line.

(3) Let a_i, i = 1,2,3, ... be the elements of the group. If two elements in ith row are equal, then $a_i a_j = a_i a_k$ for some j and some k,j ≠ k. But then, $a_i^{-1} a_i a_j = a_i^{-1} a_i a_k$, hence $a_j = a_k$. From this contradiction we deduce the required result. Similarly for columns.

(4) One group of order 2

	e	a
e	e	a
a	a	e

One group of order 3

	e	a	b
e	e	a	b
a	a	b	e
b	b	e	a

Compare these results with exercise (6) below.

Two groups of order 4

	e	a	b	c
e	e	a	b	c
a	a	b	c	e
b	b	c	e	a
c	c	e	a	b

	e	a	b	c
e	e	a	b	c
a	a	e	c	b
b	b	c	e	a
c	c	b	a	e

In the first instance you will probably obtain four apparently distinct multiplication tables. But inspection shows that, with a little rearrangement, they reduce to just two essentially distinct table, i.e. some of the groups are isomorphic; see section 10.3.

(5) The multiplication table for D_3 is:

	e	r	r^2	u_1	u_2	u_3
e	e	r	r^2	u_1	u_2	u_3
r	r	r^2	e	u_3	u_1	u_2
r^2	r^2	e	r	u_2	u_3	u_1
u_1	u_1	u_2	u_3	e	r	r^2
u_2	u_2	u_3	u_1	r^2	e	r
u_3	u_3	u_1	u_2	r	r^2	e

The product of transformations is associative. There is an identity e. Inverses are as follows:

Element	e	r	r^2	u_1	u_2	u_3
Inverse	e	r^2	r	u_1	u_2	u_3
Order	1	3	3	2	2	2

Thus D_3 is a group. It is *not* abelian.

(6) Let $g \in G$ and $g \neq e$. Let $H = \langle g \rangle$. Then $|H|$ divides $|G|$, which is a prime p. Thus $|H| = $ p. Thus $H = G$. Hence G is cyclic.

(7) Let H be a subgroup of D_3. Then $|H|$ divides $|D_3| = 6$. Thus $|H|$ must be $1, 2, 3$, or 6. By exercise (6), above, groups orders $2, 3$ must be cyclic, generated by elements of orders 2 and 3 respectively. Hence subgroups of D_3 are $\{e\}$, $\{e,u_1\}$, $\{e,u_2\}$, $\{e,u_3\}$, $\{e,r,r^2\}$, and D_3 itself.

(10) No. Consider

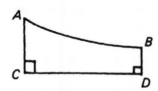

(11) Fig. 8.6.4 illustrates the technique.

(12) If OP has length l, construct P' on OP extended so that OP' has length nl. Then invert P' in the circle centre O radius l.

(13) See Fig. 8.6.5.

The vector field is reflected into those parts of the coaxial circles *inside* the disc model of the hyperbolic plane.

(15) At any vertex a segment is produced containing each angle of the polygon, in natural order. The cycle is repeated several times, completing the tessellation about that vertex.

(18) (a) Any pair of distinct lines meet at exactly one point.

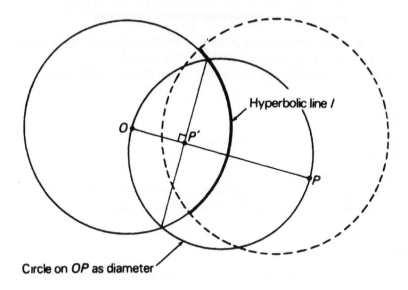

Hyperbolic line *l*

Circle on *OP* as diameter

Fig. 8.6.4.

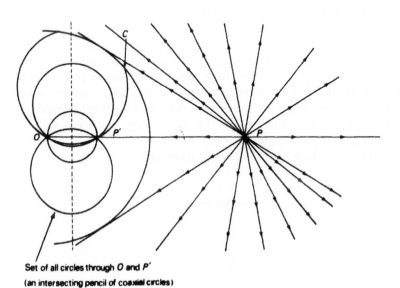

Set of all circles through *O* and *P′*
(an intersecting pencil of coaxial circles)

Fig. 8.6.5.

(b) Greater than 2π.

(c) Working as in section 8.4, we see that the possible regular tessellations are

$$\{2,q\},\ \{p,2\},\ \{3,3\},\ \{3,4\},\ \{4,3\},\ \{3,5\},\ \{5,3\}\ .$$

For $\{2,q\}$ take arcs joining a point A to the diametrically opposite point A'.

For $\{p,2\}$ take vertices and arcs around the boundary.

For the other tessellations cut models of the Platonic solids in half.

EXERCISES 9.6

(4) $a_1a_2a_3a_4a_2a_1a_2a_5a_3a_4a_3a_1$ is one example.

(6) (a) 1; (b) 2.

(7) In each case three colours suffice.

(8) The reading for the vertices suggests that 3 and 6 are labelled a and h. Now adapt the algorithm for colouring maps to this situation. (*cgadehfb*).

EXERCISE 10.5

(1) Rearrange the multiplication tables according to the correspondence $x \leftrightarrow ab$ and $y \leftrightarrow b$.

(2) To show H is isomorphic to D_3 let x correspond to r and y and u_1, and the other elements accordingly. Since, by exercise (1), above, G is isomorphic to H, it then follows that G is also isomorphic to D_3.

(3) Let g correspond to $1 \in \mathbf{Z}$. Then an arbitrary element of G can be written uniquely in the form g^r and corresponds to r. Another such element of G is g^s which corresponds to s. Then $g^r.g^s = g^{r+s}$ corresponds to $r + s$, as required. Thus the groups G and Z are isomorphic.

(The usual definition of two groups being isomorphic is: if G and H are two groups and f is a one – one and onto function from G to H such that, for *any* g_1, $g_2 \in G$, $f(g_1g_2) = f(g_1)f(g_2)$, then G and H are isomorphic groups and f is an isomorphism. A moments thought will show that this says that, with a suitable correspondence between the elements of the two groups, the multiplication tables correspond, which was our previous definition. In the above case, we have $f(g^r) = r$. See references [11] and [10] for further details.)

(4) $\pi(P) = \langle a|a^2 = e \rangle$, which has the multiplication table

	e	a
e	e	a
a	a	e

Since $\{1, -1\}$ has the same multiplication table under the correspondence $1 \leftrightarrow a^2 = e$ and $-1 \leftrightarrow a$, the two groups are isomorphic. In fact, both groups are examples of a cyclic group of order 2.

(5) $\pi(T) = \langle a,b | aba^{-1}b^{-1} = e\rangle$. This is the same as $\langle a,b | ab = ba\rangle$. We define the function f from $\mathbf{Z} \times \mathbf{Z}$ to $\pi(T)$ by $f((n,m)) = a^n b^m$. Now show that f is one – one and onto (i.e. $f(x) = f(y)$ implies that $x = y$ and, if $w \in \pi(T)$, then there exists an element $x \in \mathbf{Z} \times \mathbf{Z}$ such that $f(x) = w$.) Finally show that $f(xy) = f(x).f(y)$ for *any* elements x and y in $\mathbf{Z} \times \mathbf{Z}$.

(6) (b) (i) Edges identified in groups of 4 and 3, and vertices in groups of 6. This is a manifold.

   (ii) Edges identified in groups of 4 and 8, and all vertices identified. This is *not* a manifold.

Exercises 11.4

(1) $(a), (b), (c)$ have 2,3,1, boundary curves respectively.

By the Classification Theorem homeomorphic surfaces must have the same number of cuffs, and therefore the same number of boundary curves.

(2) The technique illustrated in section 3.3 gathers together the T's using operations 1 to 3. Operation 4 gathers together the P's as illustrated in section 3.4. Section 3.6 shows how operatons 1 to 4 are used to convert sequences of P's to the form mTP or mTK. The proof of Theorem 11.1.2 shows how operations 1 to 4 are used to gather together the cuffs.

These give the normal forms:

$$\text{(a)} \quad P\bar{D}, \quad \text{(b)} \quad T\bar{D}, \quad \text{(c)} \quad TP\bar{D}, \quad \text{(d)} \quad TK\bar{D}, \quad \text{(e)} \quad K2\bar{D}$$

These are checked by finding the number of distinct vertices v on the plane model, and therefore $\chi(M)$, then finding the number of boundary curves m, and finally matching with the appropriate formula in Lemma 11.2.1.

(a) non-orientable, $v = 2$, $\chi(M) = 0$, $m = 1$, so $M = P\bar{D}$.

(b) orientable, $v = 3$, $\chi(M) = -1$, $m = 1$, so $M = T\bar{D}$.

(c) non-orientable, $v = 2$, $\chi(M) = -2$, $m = 1$, so $M = TP\bar{D}$.

(d) non-orientable, $v = 2$, $\chi(M) = -3$, $m = 1$, so $M = TK\bar{D}$.

(e) non-orientable, $v = 3$, $\chi(M) = -2$, $m = 2$, so $M = K2\bar{D}$.

(3) (a) 3 Seifert circles, 4 crossings, so the surface has normal form $T\bar{D}$.
           [Notice this surface has the same normal form as the surface con-
           structed for the trefoil in the text. So a little more effort is needed
           if we wish to use our classification theorems for surfaces as an aid
           in distinquishing between Knots (see the Knot Theory books re-
           ferred to at the end of Chapter 11 for more details.]

     (b) 2 Seifert circles, 5 crossings, so the surface has normal form $2T\bar{D}$.

Further reading and references

[1] Appel, K.i. & Haken, W. (1976) Every planar map is four colourable, *Bulletin of the American Math. Soc.*, **82**, 711-712.

[2] Armstrong, M.A. (1979) *Basic Topology*, McGraw-Hill.

[3] Bonola, R. (1955) *Non-Euclidean Geometry*, Dover.

[4] Coxeter, H.S.M. (1969) *Introduction to Geometry*, Wiley.

[5] Coxeter, H.S.M. (1973) *Regular Polytopes*, Dover.

[6] Coxeter, H.S.M. & Moser, W.O.J. (1980) *Generators and Relations for Discrete Groups*, 4th edition, Springer-Verlag.

[7] Croom, F.H. (1978) *Basic Concepts of Algebraic Topology*, Springer-Verlag.

[8] Crowell, R.H. & Fox, R.H. (1963) *Introduction to Knot Theory*, Springer-Verlag.

[9] Escher, M.C. (1973) *The Graphic Work of M.C. Escher*, 4th printing, Pan/Ballantine.

[10] Gardiner, C.F. (1980) *A First Course in Group Theory,* Springer-Verlag.

[11] Gardiner, C.F. (1981) *Modern Algebra*, Ellis Horwood.

[12] Gardner, M. (1976) *Mathematical Carnival*, Allen & Unwin.

[13] Giblin, P.J. (1977) *Graphs, Surfaces and Homology*, Chapman & Hall.

[14] Greenberg, M.J. (1974) *Euclidean and Non-Euclidean Geometries*, W.H. Freeman.

[15] Grossman, I. & Magnus, W. (1964) *Groups and Their Graphs*, Random House.

[16] Henle, M. (1970) *A Combinatorial Introduction to Topology*, W.H. Freeman.

[17] Jeger, M. (1966) *Transformation Geometry*, Allen & Unwin.

[18] Johnson, D.L. (1980) *Topics in the Theory of Group Presentations*, Cambridge University Press.

[19] Kline, M. (1972) *Mathematical Thought from Ancient to Modern Times*, Oxford University Presss.

[20] Kosniowski, C. (1980) *A First Course in Algebraic Topology*, Cambridge University Press.

[21] Lockwood, E.H. & Maxmillan, R.H. (1978) *Geometric Symmetry*, Cambridge University Press.

[22] Lyndon, R.C. & Schupp, R.E. (1977) *Combinatorial Group Theory*, Springer-Verlag.

[23] Macdonald, I.D. (1968) *The Theory of Groups*, Oxford.

[24] Magnus, W., Karrass, A., & Solitar, D. (1976) *Combinatorial Group Theory*, Dover.

[25] Massey, W.S. (1967) *Algebraic Topology: An Introduction*, Harcourt, Brace & World.

[26] Meschkowski, H. (1965) *Evolution of Mathematical Thought*, Holden-Day.

[27] Rotman, J.J. (1973) *The Theory of Groups*, 2nd edition, Allyn & Bacon.

[28] Sawyer, W.W. (1955) *Prelude to Mathematics*, Penguin Books.

[29] Skilling, J. The complete set of uniform polyhedra, *Phil. Trans. of the Royal Soc. of London*, Series A, **278**, no. 1278.

242   **Further reading and references**

[30] Stillwell, J. (1980) *Classical Topology and Combinatorial Group Theory*, Springer-Verlag.
[31] Wenninger, M.J. (1971) *Polyhedron Models*, Cambridge University Press.
[32] Wenninger, M.J. (1979) *Spherical Models*, Cambridge University Press.
[33] Weyl, H. (1952) *Symmetry*, Princeton University Press.

Note. Although the books below are not mentioned in the text, they are recommended for further reading.
Appel, K. & Haken, W. (1989) *Every Planar Map is Four Colorable*, American Mathematical Society.
Berge, C. (1962) *The Theory of Graphs*, Methuen.
Bröcker, Th. & Jänich, K. (1982) *Introduction to Differential Topology*, Cambridge University Press.
Burghes, D.N. & Borrie, M.S. (1980) *Modelling with Differential Equations*, Ellis Horwood.
Flapan, E. (2000) *When Topology Meets Chemistry*, Cambridge University Press.
Francis, G.K. (1987) *A Topological Picture Book*, Springer-Verlag.
Gardiner, C.F. *Algebraic Structures*, Ellis Horwood.
Griffiths, H.B. (1981) *Surfaces*, 2nd edition, Cambridge University Press.
Grunbaum, B. & Shephard, G.C. (1987) *Tilings and Patterns*, W.H. Freeman.
Hall, N. (1967) *Combinatorial Theory*, Blaisdell.
Harary, F. (1969) *Graph Theory*, Addison-Wesley.
Kappraff, J. (1991) *Connections, The Geometric Bridge between Art and Science*, McGraw-Hill, Inc.
Pedeo, D. (1970) *A Course of Geometry*, Cambridge University Press.
Ryser, H.J. (1963) *Combinatorial Mathematics*, The Mathematical Association of America, Wiley.
Stewart, I. (1992) *The Problems of Mathematics*, Oxford University Press.
Sutherland, W.A. (1993) *Introduction to Metric and Topological Spaces*, Oxford University Press.
Temperley, H.N.V. (1982) *Graph Theory and Applications*, Ellis Horwood.
Weeks, J.R. (1985) *The Shape of Space*, Marcel Dekker.
Wilson, R. (1972) *Introduction to Graph Theory*, Oliver & Boyd.
Wilson, R. & Watkins, J.J. (1990) *Graphs, an introductory approach*, Wiley.

Index

Woodhead Publishing Series in Mathematics and Applications

MATHEMATICS IN TEACHING PRACTICE:
A Guide for University and College Lecturers
JOHN MASON, Institute of Educational Technology, The Open University, Milton Keynes

ISBN: 1-898563-79-9   *ca.* 200 pages   hardback   2001

This book clarifies the distinction between mathematical research and mathematical education. Intensely practical, it offers hundreds of suggestions for making small and medium sized changes for lectures, tutorials, task design or problem solving. Here is inspiration for effective mathematics teaching in a modern technological environment, directed to those now training or newly entering this profession, or unhappy with results or experiences.

Contents: Student difficulties with mathematics, statistics, modelling: techniques, concepts and logic; Lecturing: structure, employing screens, tactics, issues of interest and stimulus, repeating lectures, hand-outs, encouraging exploration and inventive exploration, support; Tutoring: atmosphere, debate, asking questions, worked examples, assent-assert, tactics, advising students, structuring tutorials; Constructing tasks: purposes, aims, intentions, summary, differing tasks and purposes, mathematics forms, objectives, challenge and routine tasks, student propensities, reproducing oneself; Marking: allocating marks, student feedback; Using history: the use of pictures, dates and potted biographies; Teaching issues and concerns: tensions, framework for informing teaching, mathematical themes;

MANIFOLD THEORY: an introduction for mathematical physicists
Daniel Martin, Department of Mathematics, Glasgow University
ISBN: 1-898563-84-5   424 pages   112 exercises: hints, solutions   2001

A authoritative account of basic manifold theory and global analysis based on advanced mathematics and physics courses and proven over many years at Glasgow University. The treatment is rigorous but "less condensed" in the sense that it offers clear and didactic explanations of parts of mathematics usually difficult to grasp.

Prequisites include a knowledge of basic linear algebra and topology. Because many courses of mathematics for physics students include a substantial amount of linear algebra but no topology, an appendix on analytic topology is included.

Plentiful worked and unworked examples appear in the text, chosen from the author's long teaching experience and his knowledge.

Contents: Vector spaces; Tensor algebra; Differentiable manifolds; Vector and tensor fields on a manifold; Exterior differential forms; Differentiation on a manifold; Pseudo-Riemannian and Riemannian manifolds; Symplectic manifolds; Lie groups; Integration on a manifold; Fibre bundles; Complex linear algebra. Almost complex manifolds; *Appendices*: Analytic topology; Quaternions and Cayley numbrs; The semidirect product of two groups; Homotopy reivew.

STOCHASTIC DIFFERENTIAL EQUATIONS & APPLICATIONS
XUERONG MAO, University of Strathclyde, Glasgow

ISBN 1-898563-26-8   360 pages   1997

This advanced undergraduate and graduate text covers basic principles and applications of various types of stochastic systems which now play an important role in many branches of science and industry. It is a source book for pure and applied mathematicians, statisticians and probabilists, and engineers in control and communications, information scientists, physicists and economists.

It emphasises the analysis of stability in stochastic modelling and illustrates the practical use of stochastic stabilisation and destabilization, stochastic oscillators, stochastic stock models and stochastic neural networks in pragmatic, real life situations.

Contents: Generalised Gronwall inequality and Bihari inequality; Introduces the Brownian motions and Stochastic integrals; Analyses the classical Ito formula and the Feynman-Kac formula; Demonstrates the manifestations of the Lyapunov method and the Ruzumikhin technique; Discusses Cauchy-Marayama's and Carathedory's approximate solutions to stochastic differential equations; and more.

EXPERIMENTAL DESIGN TECHNIQUES IN STATISTICAL PRACTICE
W.P. GARDINER, Department of Mathematics, Glasgow Caledonian University, G. GETTINBY, University of Strathclyde, Glasgow

ISBN 1-898563-35-7   256 pages   1997

Describes classical and modern design structures together with new developments which now play a significant role for quality improvement and product development, principally for interpretation of data from industrial and scientific research. The material is reinforced with software output for analysis purposes, and covers much detail of value to users, such as model building and derivation of expected mean squares.

Contents: Introduction; Inferential data analysis for simple experiments; One-factor designs; One-factor blocking designs; Factorial experimental designs; Hierarchical designs; Two-level fractional factorial designs; Two-level orthogonal arrays; Taguchi methods; Response-surface methods; Appendices: Statistical tables; Glossary; Problems and answers.

DECISION AND DISCRETE MATHEMATICS
prepared by **THE SPODE GROUP** with Ian Hardwick, Truro School, Cornwall

ISBN 1-898563-27-6   240 pages   1996

A complete coverage in the Decision Mathematics (or Discrete Mathematics) module of A-level examination syllabuses. Also suitable for first year undergraduate courses in qualitative studies or operational research, or for access courses. Reflects the combined teaching skills and experience of authors within The Spode Group. The text is modular, explaining concepts used in decision mathematics and related operational research, and electronics. Emphasises techniques and algorithms in real life situations. Clear diagrams; plentiful worked examples; Exam-standard questions; Many exercises.

Contents: Introduction to networks; Recursion; Shortest route; Dynamic programming; Network flows; Critical path analysis; Linear programming: Graphical; Linear programming: Simplex method; The transportation problem; Matching and assignment problems; Game theory; Recurrence relations; Simulation; Iterative processes; Sorting; Algorithms; Appendices; Answers; Glossary.

Choice: American College Library Association
"Topics include networks, recursive and iterative processes, critical path analysis, linear programming, the transportation problem, recurrence, game theory simulation, and sorting/packing algorithms ... explores both key concepts and fundamental algorithms, but also relates the area of decision mathematics to real-life situations ... lower-division undergraduates, two-year technical program students."

MODELLING AND MATHEMATICS EDUCATION: ICTMA 9
Applications in Science and Technology
J.F. MATOS, University of Lisbon, Portugal; **S.K. HOUSTON**, University of Ulster; **W. BLUM**, University of Kassel, Germany; **S.P. CARREIRA**, University of Lisbon, Portugal

ISBN: 1-898563-66-7   300 pages   hardback   2000

This book records the 1999 Lisbon Conference of ICTMA (The International Conference on Mathematical Modelling & Applications). It contains the selected and edited content of the conference and makes a significant contribution to mathematical modelling which is the significant investigative preliminary to all scientific and technological applications, from machinery to satellites and docking of spaceships.

Contents: Weak derivates; Problems of research on teaching and learning of applications and modelling; Enacting possible worlds: making sense of (human) nature; There is more to mathematical modelling than just producing a model; The metaphorical nature of mathematical models; The secondary school curriculum for mathematical modelling and applications.

TEACHING AND LEARNING MATHEMATICAL MODELLING:
Innovation, Investigation and Application
S.K. HOUSTON, University of Ulster, Northern Ireland; W.BLUM, University of Kassel, Germany; IAN HUNTLEY, University of Bristol, England; N.T.NEILL, University of Ulster, Northern Ireland

ISBN- 1-898563-29-2   416 pages   1997

Sponsored by the organising committee of International Conferences on the Teaching of Mathematical Modelling and Applications (ICTMA), this book contributes to teaching and learning mathematical modelling in universities throughout degree study, colleges of technology, teachers' training colleges, high schools and sixth form colleges.

Contents: Reflections and investigations; Assessment at undergraduate level; Secondary courses and case studies; Tertiary case studies; Tertiary courses.

The Mathematics Teacher (Dr Barry E. Shealy, Buffalo, USA)
"The greatest value would be for people in postsecondary contexts conceptualising courses centred around mathematical applications and modelling."

Choice: American College Library Association
"Of great value to those teaching mathematical modelling in high schools, colleges, and universities."

Zentralblatt für Mathematik und ihre Grenzgebiete, Berlin,
"Deals with assessment, particularly at undergraduate level, secondary education and case studies of good practice with examples of courses and how they are taught in a variety of countries including Russia, the Netherlands, the USA and the UK. Concludes with descriptions of ideas relating to undergraduate modelling courses."

MATHEMATICAL MODELLING:
Teaching and Assessment in a Technology-Rich World
P. GALBRAITH, University of Queensland, Brisbane, Australia; W. BLUM, The University of Kassel, Germany; G. BOOKER, Griffith University, Brisbane, Australia; IAN D .HUNTLEY, University of Bristol, England

ISBN 1-898563-42-X   368 pages   Hardback   1997

This book contributes to the teaching, learning and assessing of mathematical modelling in this era of rapidly expanding technology. It addresses all levels of education, from secondary schools through teacher training colleges, colleges of technology, universities, and state and national departments of mathematical education and research groups.

Contents: Issues and alternatives in assessing modelling; Technologically enriched mathematical modelling; Real world: Models and applications; Applications and modelling in teaching and learning; Applications and modelling in a system or national context.

GAME THEORY: Mathematical Models of Conflict
A.J. JONES

ISBN: 1-898563-14-4   *ca.* 300 pages   1999

A modern, up-to-date text for senior undergraduate and graduate students (plus teachers and professionals) of mathematics, economics, sociology; and operational research, psychology, defence and strategic studies, and war games. Engagingly written with agreeable humour, this account of game theory can be understood by non-mathematicians. It shows basic ideas of extensive form, pure and mixed strategies, the minimax theorem, non-cooperative and co-operative games, and a "first class" account of linear programming, theory and practice. The book is self-contained with comprehensive references from source material.

DELTA FUNCTIONS: A Fundamental Introduction to Generalised Functions
R.F. HOSKINS, Department of Mathematical Sciences, De Montfort University, Leicester

ISBN: 1-898563-44-6   250 pages   1999

Provides modules for advanced undergraduates and postgraduates reading applied mathematics, theoretical physics and electrical (signal processing) and mathematically-based mechanical engineering. It deals with generalised functions in an easy way, making it accessible to a broad audience of scientists and engineers. The text addresses the continual problem of the transition from elementary delta functions and its applications. The interplay between the properties and applications of generalised functions is carefully explained, and the introduction of non-standard concept is new and innovative. Little more than a standard background in calculus is assumed, and attention is focused on techniques, with a liberal selection of worked examples and exercises.

Contents: Part I: Generalised Functions: Summary results from elementary analysis; Properties of the delta function and its derivatives; Applications to linear systems theory, Laplace transforms and fourier analysis; Part II: Delta Functions: A Simple Form of NSA; Representations of delta functions as internal non-standard functions; Other types of generalised function; Distribution.

Professor H. Westcott Vayo, Mathematics Dept, University of Toledo, USA
"I found this a good book (first edition) to teach from and the students can actually read it."

Choice: American College Library Association
"By the end of the first chapter, undergraduate readers can solve simple two-person zero-sum games starting from its rules. Solutions, many details provided for all problems."

GEOMETRY OF NAVIGATION
ROY WILLIAMS, Master Mariner, BSc, PhD, FRIN, AFIMA

ISBN: 1-898563-46-2   144 pages   Hardback   1999

Contents: Geometrical representation of the earth: Mathematics of chart projections: Navigating along Rhumb lines: Shortest paths on the surface of a sphere: Shortest paths on the surface of an ellipsoid: Paths between nearly antipodean points: Great ellipse on the surface of an ellipsoid: Navigating along the arc of a small ellipse: Surface position from astronomical observation: Surface position from satellite data: Appendices: Table of latitude parts (meridian distance): Transformation between equations: Direct cubic spline approximation.

Journal of Navigation:
"Sets out [his] results in the form of 'computational procedures' which can be followed by non-specialists ... Roy Williams has undertaken a difficult, but much-needed, task in applying modern mathematical methods to both old and new navigational problems. He has met this challenge with considerable success in this well-produced and clearly presented book."

ORDINARY DIFFERENTIAL EQUATIONS & APPLICATIONS: Mathematical methods for applied mathematicians, physicists, engineers and bioscientists
W.S. WEIGELHOFER & K.A. LINDSAY, University of Glasgow

ISBN: 1-898563-57-8   224 pages   1999

This advanced undergraduate text provides the basis for a semester/module of 20-25 lectures for students of applied mathematics and the applied sciences of physics, engineering, chemistry and biology. It approaches the study of ordinary differential equations: the more traditional setting, first-order differential equations and their solutions; followed by a practical and modern approach to mathematical modelling. It emphasises how various types of differential equations arise in simple modelling scenarios in applied mathematics and physics, the engineering and biological sciences.

Contents: Introduction and revision; Modelling applications; Linear differential equations of second order; Oscillatory motion; Miscellaneous solution techniques; Lapalace transform; High order initial value problems; System of first order linear equations; Boundary value problems; Optimization; Calculus variation; Appendix A: Self study projects; Appendix B: Extended tutorial solutions.

SIGNAL PROCESSING IN ELECTRONIC COMMUNICATION
MICHAEL J. CHAPMAN, DAVID P. GOODALL, and NIGEL C. STEELE, School of Mathematical and Information Sciences, University of Coventry

ISBN 1-898563-30-6   288 pages   1997

Contents: Signal and linear system fundamentals; System responses; Fourier methods; Analogue filters; Discrete-time signals and systems; Discrete-time system Responses; Discrete-time Fourier analysis; The design of digital filters; Aspects of speech processing; Appendices: The complex exponential; Linear predictive coding algorithms; Answers.

LINEAR DIFFERENTIAL AND DIFFERENCE EQUATIONS:
A systems approach for mathematicians and engineers
R.M. JOHNSON, Department of Mathematics and Statistics, University of Paisley

ISBN: 1-898563-12-8   200 pages   1996

This text for advanced undergraduates and graduates reading applied mathematics, electrical, mechanical, or control engineering employs block diagram notation to highlight comparable features of linear differential and difference equations, a unique feature found in no other book. The treatment of transform theory (Laplace transforms and z-transforms) encourages readers to think in terms of transfer functions, i.e., algebra rather than calculus. This contrives short-cuts whereby steady-state and transient solutions are determined from simple operations on the transfer functions.

Institute of Electrical Engineers (IEE) Proceedings
"Should find wide application by undergraduate students in engineering and computer science ... the author is to be congratulated on the importance that he attaches to conveying the parallelism of continuous and discrete systems."

MATHEMATICAL KALEIDOSCOPE: Applications in Industry, Business and Science
BRIAN CONOLLY, Emeritus Professor of Mathematics, (Operational Research), University of London *and*
STEVEN VAJDA, Sussex University, *formerly* Professor of Operational Research, University of Birmingham

ISBN: 1-898563-21-7   276 pages   1995

Contents: Miscellaneous fantasies; Finance; Games; Mathematical programming; Search, pursuit, rational outguessing; Organisation and management; Mathematical teasers; Triangular geometry.

Financial Risk in Insurance:
"A wide variety of topics, all related to applied mathematics. Readership is advanced undergraduates and postgraduates in applied mathematics, statistics and operational research; researchers and applied mathematicians in professional practice; and careers officers."

London Mathematical Society Newsletter:
"Precise, sometimes amusing, reflecting the unusual background of the topics ... original in essence or presentation. Since the authors are skilful in manipulating mathematical expressions, the reader cannot escape from vigorous mental exercise!"

Journal of the Royal Statistical Society:
"Here we have two very experienced players describing aspects they have enjoyed. It is light-hearted and helps to transmit the enthusiasm of the authors."

MATHEMATICAL ANALYSIS AND PROOF
DAVID S. G. STIRLING, Senior Lecturer in Mathematics, University of Reading

ISBN 1-898563-36-5   256 pages   1997

Contents: Setting the scene; Logic and deduction; Mathematical induction; Sets and numbers; Order and inequalities; Decimals; Limits; Infinite series; Structure of real number system; Continuity; Differentiation; Functions defined by power series; Integration; Functions of several variables.

The Mathematical Gazette:
"A fine line between accuracy and exactitude. David Stirling treads it carefully. A thorough a comprehensive introduction. Very much in the classical mould but written in a chatty style with the common student misunderstandings in mind. It should be in your undergraduate reference library."

Choice: American College Library Association, USA
"Get down to serious analysis at the level of Walter Rudin. Standard analytic topics are treated. The level is roughly that of Bartle, and Stirling makes an effort to explain the plans of attack for certain proofs as well as presenting them."

Mathematics Today:
"The book can be said to be self-contained. One of the better ones I have seen to date. I will definitely be recommending the text without hesitation and encourage other lecturers to give it serious consideration as a potential teaching aid."

FUNDAMENTALS OF UNIVERSITY MATHEMATICS, Second edition
COLIN McGREGOR, JOHN NIMMO and WILSON W. STOTHERS, Department of Mathematics, University of Glasgow

ISBN: 1-898563-   540 pages   300 worked examples   750 exercises Hardback   2000
A unified course for first year mathematics, bridging the school/university gap, suitable for pure and applied mathematics courses, and those leading to degrees in physics, chemical physics, computing science, or statistics. The treatment is careful, thorough and unusually clear, and the slant and terminology are modern, fresh and original, in parts sophisticated and demanding some student commitment. Fresh ideas for teachers, students, and tutorials. 300 worked examples, rigorous proofs for most theorems, 750 exercises with answers. Problems and solutions for all topics.

Contents: Preliminaries, functions & inverse functions; Polynomials & rational functions; Induction and the binomial theorem; Trigonometry; Complex numbers; Limits and continuity; Differentiation - fundamentals; Differentiation - applications; Curve sketching; Matrices & linear equations; Vectors & three dimensional geometry; Products of vectors; Integration - fundamentals; Logarithms & exponentials; Integration - methods and applications; Ordinary differential equations; Sequences & series; Numerical methods.

The Mathematical Gazette:
"The borderline between school and university mathematics is always a tricky one. This book sits firmly on the university side of the borderline. If you are looking for a first year university text you should certainly look at this one."

MATHEMATICS FOR EARTH SCIENCES
PATRICK SHARKEY, School of Mathematics Studies, University of Portsmouth

ISBN: 1-898563-62-4   280 pages   1999

This text covers the relevant mathematics for courses in geology (including environmental), applied mathematics, geography, geophysics, palaeobiology, engineering geology (soil mechanics, construction, hydraulics), mining (exhaustible resource like oil etc), and surveying. The approach is deliberately gentle and intuitive for non-mathematical scientists.

Rich in worked examples, the book addresses the mathematical concepts as they apply to the geological sciences in the wake of the great upheavals caused by tectonics and other phenomena. Each chapter concludes with a summary of its content, problem exercises, answers, and tutorial help from teaching experience in UK and USA.

Contents: Introduction (brief history, angles, lengths); Vectors and Matrices (Markhov Processes, Elasticity); Plane geometry (Surveyor's Formula, Dip and Strike); Solids (Mensuration, Symmetry, Crystal Shapes); Spheres (Geodesics, Projections, Co-ordinates, Spherical Trigonometry); Elmentary Calculus (Rates of Change, Curvature, Planetary Orbits); Two-Variable Calculus (Landforms, Gradients, Errors, Lava Flow); Differential and Difference Equations (Seismometers, River Meander, Radioactivity); Probability (Elementary Ideas in Decision Making); Risk Analysis (Applications to Explorations and Earthquake Occurrence); Optimisation (Elementary Ideas from Variational Calculus and Optimal , Exhaustible Resources.

CALCULUS
Introduction to Theory and Applications in Physical and Life Science
R.M. JOHNSON, Department of Mathematics and Statistics, University of Paisley

ISBN: 1-898563-06-3   336 pages   1995

This lucid and balanced introduction for first year engineers and applied mathematicians conveys the clear understanding of the fundamentals and applications of calculus, as a prelude to studying more advanced functions. Short and fundamental diagnostic exercises at chapter ends testing comprehension, before moving to new material.

Contents: Prerequisites from algebra, geometry and trigonometry; Limits and differentiation; Differentiation of products and quotients; Higher-order derivatives; Integration; Definite integrals; Stationary points and points of inflexion; Applications of the function of a function rule; The exponential, logarithmic and hyperbolic functions; Methods of integration; Further applications of integration; Approximate integration; Infinite series; Differential equations.

Dr John Baylos, Faculty of Science and Mathematics, Nottingham Trent University:
"I have decided to use this book as a core text for a basic module in the first year of BSc/HND engineering courses. Many students have a fairly weak mathematical background, and this text is suited in pace, content, range of well-chosen examples, for students needing a high level of support."

THERMODYNAMICS OF EXTREMES
BERNARD H. LAVENDA, Research Centre for Thermodynamics, National Laboratory for Technology, Energies and Environment, Cassaccia, Rome *and* University of Camerino, Italy

ISBN 1-898563-24-1   256 pages   1995

This book offers an alternative to traditional thermodynamics, strengthening the role of entropy as a bridge between probabilistic foundations of thermodynamics and macroscopic thermodynamics. It revitalises the extreme value theory, and adds the hitherto "lacking" thermodynamic concepts which unveil its power of analysis and generality in astronomy, chemical physics, cosmology, spectroscopy, low-temperature physics, polymer and materials science, special and general relativity, structural engineering and design.

Contents: Why thermodynamics of extremes? Fundamental principles of thermo-dynamics; Phenomenology of extremes; Thermostatistics of polymer chains; Cosmology and thermogravity; Thermostatistics of materials.

Choice: American College Library Association
"Vigorously pushes thermodynamic theories to their limits and covers a very wide range of physical effects, presented mostly as a collection of equations with topics from "hot" degenerate stars to Landau's theory of superfluidity to 'black radiation'. For careful specialists who want to challenge other thermodynamicists."

J. Dunning-Davies, University of Hull
"Sets the volume apart from other books with thermodynamics of degenerate systems treated in great detail. New, interesting results emerge. An unusual thermodynamics book containing novel ideas and solutions to several well-known problems which should be found in every respectable thermodynamics library."

CONCISE THERMODYNAMICS: Principles and Applications
J. DUNNING-DAVIES, Senior Lecturer in Applied Mathematics, University of Hull

ISBN: 1-85463-15-2   200 pages   1996

Contents: .Introduction; The Zeroth law; The first law; The second law; Non-stochastic processes; The third law; Extension to open and non-equilibrium systems; Thermo-dynamic cycles; Negative temperatures and the second law; Phase transitions; Thermo-dynamic equilibrium and stability; Concavity of the entropy and negative heat capacities; Black hole entropy and alternative model for a black hole; Concluding remarks; Appendix; Answers and solutions to problems.

Professor Peter T. Landsberg, University of Southampton
"A widely used discipline. This pleasant and slim volume offers a rapid and simple introduction which can be used by physicists, chemists and engineers. There are problems with solution which will help the beginner to sort out his ideas."

ENGINEERING THERMODYNAMICS
Gas and Steam Cycles for Converting Heat into Work
GEORGE COLE, Engineering Design & Manufacture Department, Hull University

ISBN 1-898563-22-5   144 pages   1996

Explains concepts behind the laws of thermodynamics for university undergraduates and graduates, or for industrialists who provide 98% of the world's electricity. Its importance lies in the conservation of energy resources while expanding work available to the world. Professor Cole demonstrates unity between gas and steam systems for energy-work conversion. Modifications and improvements of the thermo-dynamic cycles are shown as consequences of the approach to the cycles.

Contents: Sources for energy; Thermodynamics background; Gases as working fluids; Steam as working fluid; Gas-steam synthesis; Applications involving combinations of cycles; Summary and future prospects.

Dr J. Dunning-Davies, University of Hull
"A welcome addition to the library of anyone interested in the provision of energy. At the heart of the global provision of work, important for anyone studying engineering or physics. Should be *compulsory* reading for physical scientists, engineers and geographers. Excellent, lucidly written; useful information and insight in a wide range of academic disciplines."

NETWORKS OF ELECTRONIC COMMUNICATIONS
R.H. JONES, School of Mathematics and Information Sciences, University of Coventry

ISBN: 1-898563-23-3   160 pages   1997

A mathematical account of important topics in communications engineering, discusses design and analysis of applications in applied graph theory. It is a course text for advanced undergraduates and postgraduates in applied mathematics, electronics, communications and computing, also workers in industrial and academic research. There is much valuable material on operational research and discrete mathematics. The author identifies problems, and then indicates the algorithms available for their solution.

Printed and bound by CPI Group (UK) Ltd, Croydon, CR0 4YY

03/10/2024

01040339-0014